建筑立场系列丛书 No.58

C3

灵活的学习空间
Learning in Fluid

汉英对照
（韩语版第374期）

韩国C3出版公社 | 编
孙探春 于风军 杜丹 徐雨晨 | 译

大连理工大学出版社

灵活的学习空间

全新的大学建筑类型

户外房间和室内广场

围合社区

C3 建筑立场系列丛书 No. 58

Learning in Fluid

A New University Building Typology

Outdoor Rooms and Indoor Squares

Enveloping Communities

灵活的学习空间

Learning i

大学建筑反映了特定时期、特定区域的教育理念。柏拉图学院、中世纪的大学、美国早期的校园和战后的大学都是如此。虽然这些大学的规划随着历史的不断积极发展发生了巨大的改变，但某些方面还是保留了下来。其中的一个方面就是教师的核心作用，这反映在学术建筑空间中教学与举办讲座时的空间的重要性中。然而，进入21世纪后这方面开始有了改变。随着知识经济、新技术、灵活的教育方法的出现，学生能更灵活地选择在哪里、以什么方式、什么时候学习。学生成为了享有特权的客户，因为学校要迎合他们来获得资金和地位。大学已经通过提供宽敞灵活的空间来体现以学生为中心的学习方式，这往往体现在大学建筑中自然元素的展现、室外自然风光的视野以及大学草坪环境的模拟中。

University buildings reflect the educational philosophies of a given geographical area during a particular period in time. This relation is true of Plato's Academy, the medieval universities, the early American campuses, and post-war universities. While the approach to university planning has differed greatly throughout these successive historical phases, certain aspects have remained consistent. One such aspect is the central role of the teacher, which is reflected in the importance of teaching and lecture spaces in academic buildings. In the 21st century, however, this aspect of education is beginning to change. With the advent of the knowledge economy, new technologies, and flexible educational programmes, students have more choice and control over where, how, and when they learn, making them privileged customers to which universities cater as they compete for status and funds. Universities have responded by providing large, flexible spaces for active student-centered learning, often featuring natural elements or views to nature, simulating the environment of a college lawn.

n Fluid

灵活的学习空间：全新的大学建筑类型

学习者的社区

大学是一个把有着共同目标的不同个体聚集在一起的地方，其目标就是创造与传播知识。换言之，就是一个学习者的社区。其结构源于成员之间的关系：即学生、老师和管理者之间的关系。这些关系的本质由特定的教育理念形成，进而表现在大学的建筑形式上。纵观历史，大学的建筑外形随着不断改变的教育理念而改变。然而，尽管已被物化为不同的方式，某些空间格局却仍然保留了下来。其中一个就是大学作为创造和传播知识的教学空间的首要地位没变，以反映教师角色在教育中的历史重要性。

这种模式可以追溯到柏拉图学院。据说柏拉图学院标志着人类有意识地考查自己的知识的开始[1]。柏拉图早期在自家收学生，学生在这种社会环境内讨论和辩论问题。柏拉图提出了提问法，鼓励不同意见者开展对话，其目的是发现真理[2]。虽然柏拉图的教学没有传授人们所认为的知识和真理的教学大纲，但他的指导对知识的探索无疑是至关重要的。

在12世纪，随着古希腊文本重新被发现，大学开始兴起。在中世纪的大学中，教学变得比柏拉图学院时期的教学更加直接。这种直接的教学方式体现在博洛尼亚大学的戴宫上。它是第一所以目标为导向的大学建筑，所有教学楼都围绕着一个庭院[3]设置。另一方面，牛津的第一座大学建筑物是学院建筑，呈四边形，最开始是用作学生宿舍，由一名教师负责管理。后来，这座建筑具有了教学功能，使学生与教授之间保持紧密的关系。17世纪清教徒殖民时期，大学模式传入美国时又遭

Flexible Learning Spaces: A New University Building Typology

A Community of Learners

A University is a corporation or association of individuals brought together for the common purpose of creating and disseminating knowledge. In other words, it is a community of learners. Its structure derives from the relations between its members: between students, teachers, and administrators. The nature of these relations is informed by a particular educational philosophy, and finds expression in the university's built form. Throughout history, universities' built forms have evolved in tandem with changing educational philosophies. However, certain spatial patterns have persisted, even if they have been materialized in different ways. One such pattern is the primacy of teaching spaces for the creation and dissemination of knowledge, reflecting the historical importance of the teacher's role in education.

This pattern can be traced back to Plato's Academy, which has been said to mark the beginning of the deliberate examination of knowledge.[1] Plato received students in his house for discussion and argument in a social context, proposing a method of enquiry based on dialogues between opposing views, with the aim of discovering the truth[2]. While Plato did not teach a syllabus of accepted knowledge and facts, his guidance was undoubtedly central to the search for knowledge.

In the medieval universities that emerged in the 12th century with the rediscovery of ancient Greek texts, teaching became more direct than it had been in Plato's Academy. This emphasis is reflected in the University of Bologna's Palazzo dell'Archiginnasio, one of the first purpose-built university buildings, which arranged teaching halls around a courtyard[3]. At Oxford, on the other hand, the first university building was the college, which took the form of a quadrangle, and originally functioned as a student boarding house presided over by a master. As it eventually took on teaching functions,

线图提供：©JCY Architects and Urban Designers

遇了一系列的改变，最终使大学变成了大学校园。美国最开始的大学建筑都是散落在一处开放景观内的独立建筑物，后来采取的美术轴向规划方式将不断增长、日益复杂的大学社区[4]整齐划一，融为一体。托马斯·杰斐逊设计的弗吉尼亚大学对美国校园模式的演变有着巨大影响。五栋建筑错落有致地分布在宽敞的大型绿化空间“草坪”两侧，通过人行步道连为一体。每座建筑的顶层有一个教授房间，下面一层有一间教室。这种模式体现了杰斐逊的“大学村”的思想，他认为紧密的师生关系非常重要。中央“草坪”由建筑物围合而成，一端面向弗吉尼亚州种植园的优美景色开放，另一端与圆形建筑图书馆[5]连接。

在20世纪初和战后时期[6]，大学社区不断扩张，因为杰斐逊设计的大学社区的布局非常清晰从而成为世界范围内普遍采用的大学校园模式。尤其是20世纪60年代和70年代，随着学生数量急剧增长，为了满足在这些远离城市的大学绿色校区的学生住宿的需求，大学规划和实验日益向城市总体规划方向发展，或向大学“城”这一社会工程方向发展。

从以教学为中心到以学生为中心的转变

在21世纪，大学再次经历快速的发展，但这次发展变化主要体现在构成大学这一整体的个体之间关系的变化中。过去的教育基于讲座模式，现在学生在选择学习方式方面发挥越来越大的作用。

知识经济的到来使政府在大学投入的财政资金大量增加，但同时也给学生上大学增加了经济压力，迫使他们在选择大学时更加精挑细选。大学为其地位和资金而竞争，学生就变成了这些知识产业——大学必须迎合的享有特权的客户。

it maintained close ties between students and professors. When the collegiate model was exported to America during the 17th century Puritan colonization, it underwent successive changes that eventually transformed it into the university campus. While the first American university buildings were separate pavilions in an open landscape, they later adopted a Beaux-Arts axial planning approach as a way of unifying the growing and increasingly complex academic communities.[4] Seminal to the evolution of the American campus model was Thomas Jefferson's design for the University of Virginia. A series of five pavilions were arranged on either side of a large green space called "the lawn", and were connected by covered walkways. Each pavilion had a professor's residence on the upper floor, and a classroom below. This model embodied Jefferson's ideal for an "academic village", and expressed his belief in the importance of close student-teacher relationships. The lawn framed by the pavilions was open on one end to a view over the Virginia plantations, and terminated at the other end with the Rotunda Library[5].

The organizational clarity of Jefferson's design for an academic community made it into a model for university campuses worldwide during the expansion of universities in the early 20th century and in the post-war years.[6] Especially during the 60s and 70s, with steeply rising student enrollment, and the need to provide accommodation on out-of-the-city greenfield campuses, the scope of university planning and experimentation increasingly grew towards urban masterplanning, or the social engineering of academic "cities".

From Teaching-Centered to Student-Centered Education

In the 21st century, universities are again experiencing rapid growth, but now with even more changes at the level of relations between the individuals that make up the corporation of the university. While in the past education was based on a lecture format, today students are playing an increasingly active role in choosing how they learn.

The advent of the knowledge economy has brought with it an exponential increase in government spending on universi-

照片提供：©James Florio

Anacleto Angelini UC创新中心，圣迭戈
Anacleto Angelini UC Innovation Center, Santiago

除了知识经济，技术变得日益重要，也在以学生为中心的教育转向中发挥了重要作用。学生现在已经可以通过远程学习获取教学资料，学习网络课程，共享课程资源，老师也只是这些资源当中的一部分。同时，专业领域内的迅猛变化意味着在大学学的知识也会迅速过时。这就要求学生终生学习，要求高校提供相应的学习机会。

提供小组学习的空间也变得日益重要。合作对跨学科研究是必不可少的，可以解决需要多个学科的综合方法来解决的复杂问题或研究发展，如生物信息学、纳米技术和可持续发展这些新兴领域。

为应对上述变化，教育也需要不断变革，以使其更加灵活。这些变化包括兼职学习、远程学习、可报名参加从全日制课程大纲发展而来的的短期课程。同时，博洛尼课课程所实施的学分累积及学分转移制度使学生能够在各种不同教育机构（大学）内深造。

全新的大学建筑类型学

随着教育项目的改变，教学楼也不断改变以适应其灵活性。一种新的建筑类型正在出现，其特点是宽敞灵活的空间，以满足学习需求的多元化。

Anacleto Angelini UC创新中心是智利天主教大学圣华校区的一部分，其建筑设计正体现了这一新类型。作为公司、企业和研究人员之间进行知识交流的中心，其空间布局以面到面交流接触对于创造知识的重要性为基础。这栋十一层的建筑物的流通核心是一个大概面积为10m×15m全高中庭，四周镶嵌在深色木框里的玻璃墙使人们可以直接看到幕墙后面的办公空间。同时，建筑物统一的混凝土立面被掏空了几个三层楼高的凉廊或"空中广场"，人们可以在此驻足进行社交。中庭的一楼设有长椅，乘坐电梯的人可以在此交流。

ties, but at the same time it has placed increasing financial pressure on students to fund their education, forcing them to be more selective in choosing an academic institution. As universities compete for status and funds, students are becoming privileged customers to which knowledge businesses – the universities – cater.

In addition to the knowledge economy, the increasing importance of technology has also played an important role in this shift towards student-centered education. Students now have access to material teaching in the form of long-distance learning, course lectures that can be viewed on the web, and course material, making the teacher just one among these resources. As well, the rapidly changing professional world means that the skills gained at university quickly become obsolete. This situation requires students to become lifelong learners, and requires universities to accommodate corresponding learning possibilities.

The provision of spaces for group work is also gaining importance. Collaboration is essential for interdisciplinary research, which addresses complex problems or research developments that require the combined approaches of multiple disciplines, such as the emerging fields of bioinformatics, nanotechnology and sustainable development.

In response to these changes, educational programmes have evolved to allow for more flexibility. Changes include the possibility of studying part-time, distance learning, and enrolling in short courses often deriving from the full-time course syllabus. As well, the system of credit accumulation and transfer that was put into place by the Bologna Process enables students to gain their education from a variety of educational institutions.

A New University Building Typology

In tandem with the changing educational programme, academic buildings have adapted to accommodate the new flexibility. A new typology of academic building is emerging, characterized by large flexible spaces that cater to a diverse range of learning needs.

线图提供：©JCY Architects and Urban Designers

照片提供：©Peter Bennetts

Ngoolark埃迪斯科文大学学生服务大楼，澳大利亚
Ngoolark, ECU Student Services Building, Australia

照片提供：©Luke Hayes

牛津大学中东研究教学中心Investcorp大楼，英国
The Investcorp Building for Oxford University's Middle East Center, UK

墨尔本大学的墨尔本设计学院，其教学空间、科研空间和工作室也同样围绕着一个大型"工作室大堂"排列布置。各个空间由走廊连接，走廊里摆放有桌椅长凳，使其成为拓展的工作空间，人们在此可以俯瞰工作室大堂。走廊下面，也就是在工作室大堂那一层，可旋转的木板墙使工作室的空间延展到大堂。悬于天花板上的如同采光井的木质体量富有雕塑感，使整栋建筑更显轻盈灵活。这里为使用者提供了学习空间，与走廊连为一体。工作室大堂温暖，明亮，提供了非正式的空间；木地板上灵活地摆放着桌椅，供人们休息、交流。镶嵌装饰的方格木屋顶使整个区域的光线自然而柔和。

牛津大学的中东研究教学中心的Investcorp大楼通过提供不同程度的开放性和灵活性的空间来应对不断变化的学习方法。一楼咖啡厅和画廊的外立面为弧线形的玻璃幕墙，使室外花园一览无余。花园里有颗古老的红杉树，建筑围绕红杉树而建。二楼阅览室的弧形白色墙壁和天花板上25个泪滴形状的天窗营造出了明亮自然的工作空间。在地下一层，内部采用橡木装饰的拥有117个座位的礼堂形成了另一处讲座空间。

埃克塞特大学的论坛是以学生为本的学习中心，将原来的图书馆和大礼堂连接一体。柱廊式室内商业街或"绿色通道"设计为学生们提供了灵活的学习空间。树木和其他街道式元素，如长椅和路灯，使得这一空间富有户外特色。商店和咖啡馆进一步强调了论坛这一建筑作为人们社交场所的功能，而波浪状的木屋顶使人们想到了周围连绵起伏的群山景致。

无独有偶，Ngoolark学生服务大楼也采用了"景观式"设计方法。Ngoolark学生服务大楼位于校园各个路径的汇合处，地势高低不平，为

The Anacleto Angelini UC Innovation Center, part of the San Joaquin Campus of the Catholic University of Chile, exemplifies this new typology. Designed as a centre for the exchange of knowledge between companies, businesses and researchers, its spatial organization is based on the importance of face-to-face contact for knowledge creation. The circulation core of the 11-story building is a roughly 10x15m full-height atrium, its glazed walls with deep wood frames giving a direct view to the office spaces behind. As well, the building's monolithic concrete facade is punctured with three-story-high loggias or "elevated squares" for social exchange. The atrium's ground floor is furnished with benches, inviting those who use the elevators to engage in conversation.

The teaching, research and studio rooms of the Melbourne School of Design at University of Melbourne are similarly arranged around a large "Studio Hall". They are connected by walkways lined with tables and benches, which can become expanded working spaces that look down onto the hall. Below these walkways, on the Studio Hall level, a wall of revolving wood panels offers the possibility to extend the studio spaces into the hall. Additional flexibility is provided by a sculptural wooden volume that hangs from the ceiling like a light well, containing study spaces that can be accessed from the walkways. The Studio Hall itself has a warm, bright, informal character; it has a wooden floor and flexibly arranged sitting areas and tables, with a coffered timber roof that bathes the space in natural light.

The Investcorp Building for Oxford University's Middle East Center addresses the changing approaches to learning by providing spaces with varying degrees of openness and flexibility. The ground floor cafe and gallery have a curved glazed facade that looks onto a garden with an old Sequoia tree around which the building bends. The first floor reading room's curving white walls and ceiling with 25 teardrop-shaped skylights create a bright, natrally lit space for working. Below ground, an oak-lined 117-seat auditorium provides additional lecture space.

The Forum at University of Exeter is a student-centered hub

埃克塞特大学论坛，德文郡，英国
The Forum at University of Exeter, Devon, UK

设计带来困难。现在，Ngoolark学生服务大楼设计将图书馆、两幢主要行政楼与校园的其他部分连为一体。该建筑底部两层的公共空间景观由较低一层的"论坛"和上面较高一层的"墩座墙公共空间组成。"论坛"层的设计意在将来用作市场，遍布如岩石般的长凳和树木，地面镶嵌的石头图案如同天然流水冲蚀一般，让人联想起"水上月光图案"。这两个室外公共场所层与主建筑的中庭相连。中庭的作用如同垂直的脊形结构，将校园各个不同的交通路径连为一体，成为学生和公众社交聚会之所。

这些大学建筑展示了当代教育空间的设计理念。与以前把大学建筑物围绕"大学草坪"布局的设计相比，如今的建筑设计将两者融为一体。在当代大学建筑设计中，开放的凉廊或露台形式的草坪已成为大学建筑物的一部分，其超大灵活的空间或将自然景观尽收眼底，或本身模拟自然景观或森林。这些室内化的"大学草坪"对新一代的学生来说具有非正式、休闲的特色，标志着从以老师为中心向以学生为中心的教育理念的转换。

that connects the existing library and Great Hall, providing flexible study spaces arranged around a galleried, indoor high street or "green passage". Lined with trees and street-like elements such as benches and lampposts, the street acquires an exterior character. Shops and cafes emphasize the Forum's role as a meeting place, and the undulating timber gridshell roof evokes the rolling hills of the surrounding landscape.

In a similar way, the design for the Ngoolark Student Services Building adopts a "landscape" approach. Located at the confluence of various campus paths, it negotiates a level change that had previously been a barrier, and connects the library and two primary administrative buildings with the rest of the campus. The landscape of public space around the building on the lower two "ground" levels consists of a lower "forum" level and an upper "podium public space". The "forum" level, designed to host a marketplace in the future, is populated with rock-like benches and trees, and has a ground of inlaid stone in a pattern of natural water flow, evoking "the pattern of moonlight on water". These two exterior public levels are connected to an atrium in the main building, which acts as a vertical spine that links the various campus circulation paths, becoming a meeting place for both students and public.

These university buildings exemplify the contemporary approach to the design of educational spaces. Compared to previous models that arranged academic buildings around a "college lawn", today's approach merges the two. The lawn has become part of the academic building, taking the form of open loggias or terraces, large flexible spaces with a view to nature, or spaces that simulate the experience of a landscape or forest. These interiorized "college lawns" have an informal, leisurely character for a new privileged generation of students, marking the recent shift from a teaching-centered to student-centered educational philosophy. Isabel Potworowski

照片提供：©Wilkinson Eyre Architects

1. Martin Pearce, *University Builders*, Chichester: Wiley-Academy, 2001, p.7
2. ibidem, p.8
3. Jonathan Coulson, Paul Roberts, Isabelle Taylor, *University Planning and Architecture: The search for perfection*, Abingdon: Routledge, 2011, p.2
4. ibidem, p.8~14
5. ibidem, p.11~13
6. Martin Pearce, *University Builders*, Chichester: Wiley-Academy, p.11

埃克塞特大学论坛

Wilkinson Eyre Architects

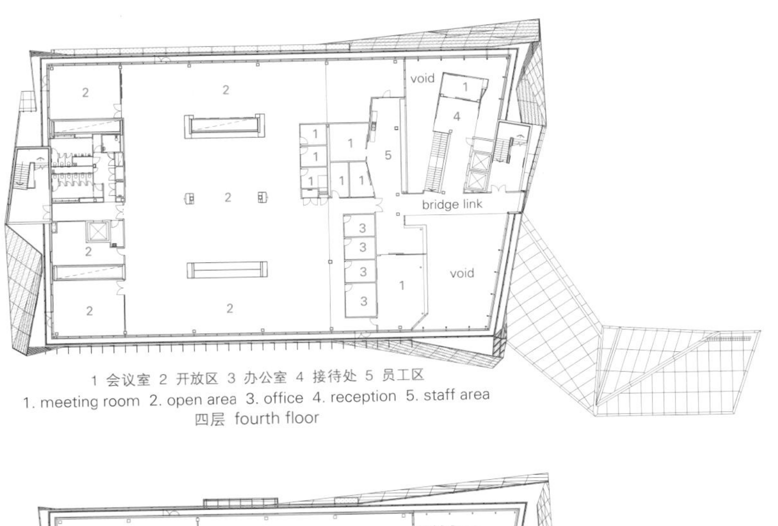

1 会议室 2 开放区 3 办公室 4 接待处 5 员工区
1. meeting room 2. open area 3. office 4. reception 5. staff area
四层 fourth floor

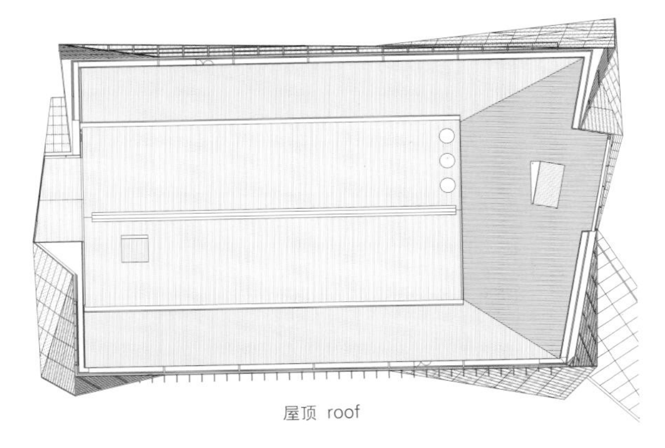

屋顶 roof

1 会议室 2 开放区 3 办公室 4 接待处 5 商店 6 员工区
1. meeting room 2. open area 3. office 4. reception 5. store 6. staff area
三层 third floor

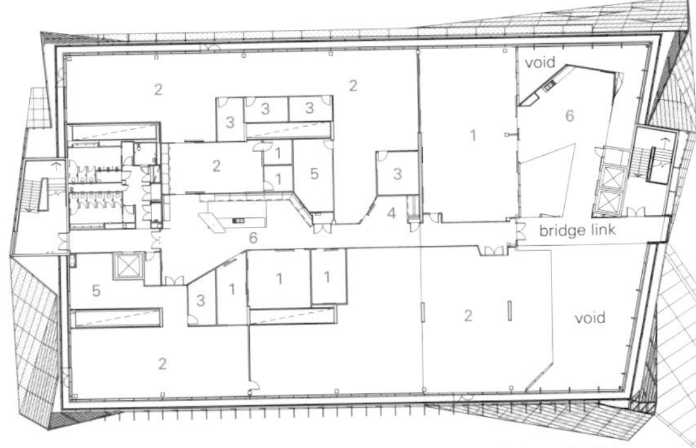

1 会议室 2 开放区 3 办公室 4 接待处 5 商店 6 员工区
1. meeting room 2. open area 3. office 4. reception 5. store 6. staff area
五层 fifth floor

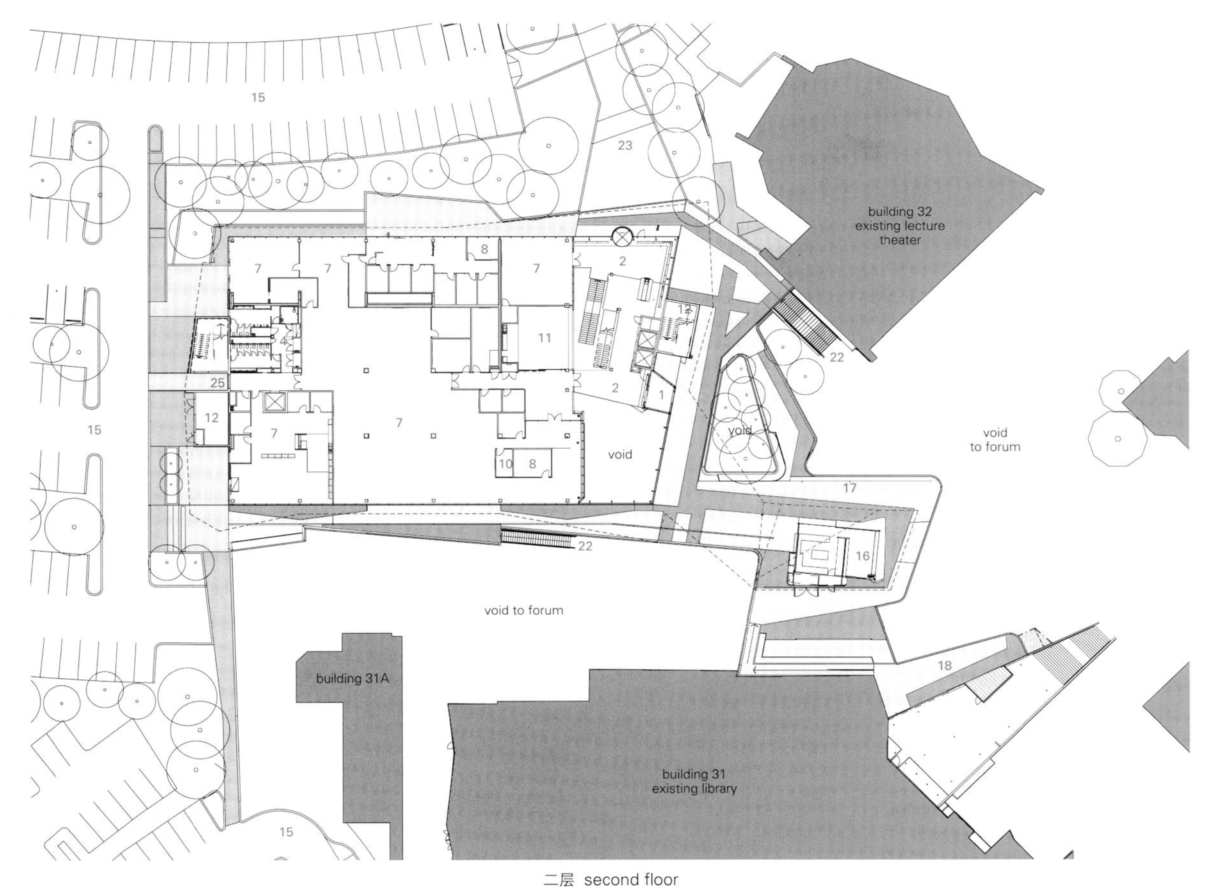

二层 second floor

1 气闸 2 中庭的公共门厅 3 讲室 4 卫生间 5 家长会客室 6 会议室 7 开放区域 8 办公室 9 接待处
10 商店 11 员工室 12 发电室 13 电梯 14 品牌游击店 15 停车场 16 咖啡室制作区 17 咖啡室的坐椅区 18 墩座墙/阳台
19 咖啡室 20 Kurongkurl Katit Jin庭院 21 Joondal公共庭院 22 下至一层的论坛空间 23 二层的北入口 24 主入口 25 服务入口

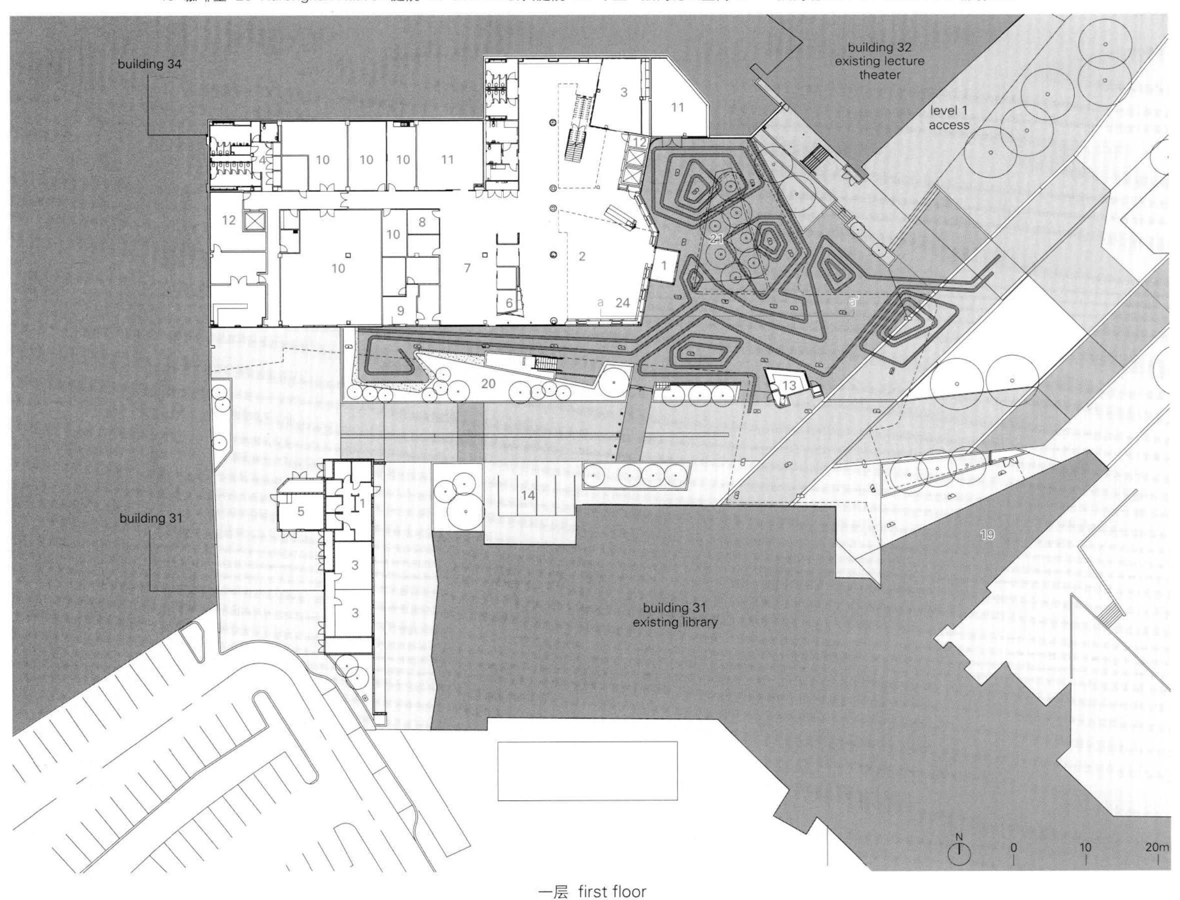

一层 first floor

1. airlock 2. atrium public foyer 3. seminar room 4. toilet block 5. parents' room 6. meeting room 7. open area 8. office 9. reception
10. store 11. staff area 12. plant room 13. lift 14. pop up shop 15. parking 16. cafe kitchen 17. cafe seating 18. podium/balcony 19. cafe
20. Kurongkurl Katit Jin courtyard 21. Joondal public courtyard 22. down to level 1 forum 23. level 2 north entry 24. main entry 25. service entry

Ngoolark, ECU Student Services Building

"Ngoolark" (the Noongar name for the Carnaby Cockatoo) brings ECU a building and place with a new living and active platform for "University Life", creating an interactive and integrated campus building bringing together the multitude of functions required to provide the highest quality of student services for the campus community and future masterplan.

The project comprises the development of a new campus building, recognizing the need to invest in the urban life of the University creating a combined campus marketplace, podium and forum.

Levels one and two provide a strong vibrant and active student hub, with the more "private" corporate structure on upper levels providing flexible and high quality office, research and innovation workplaces.

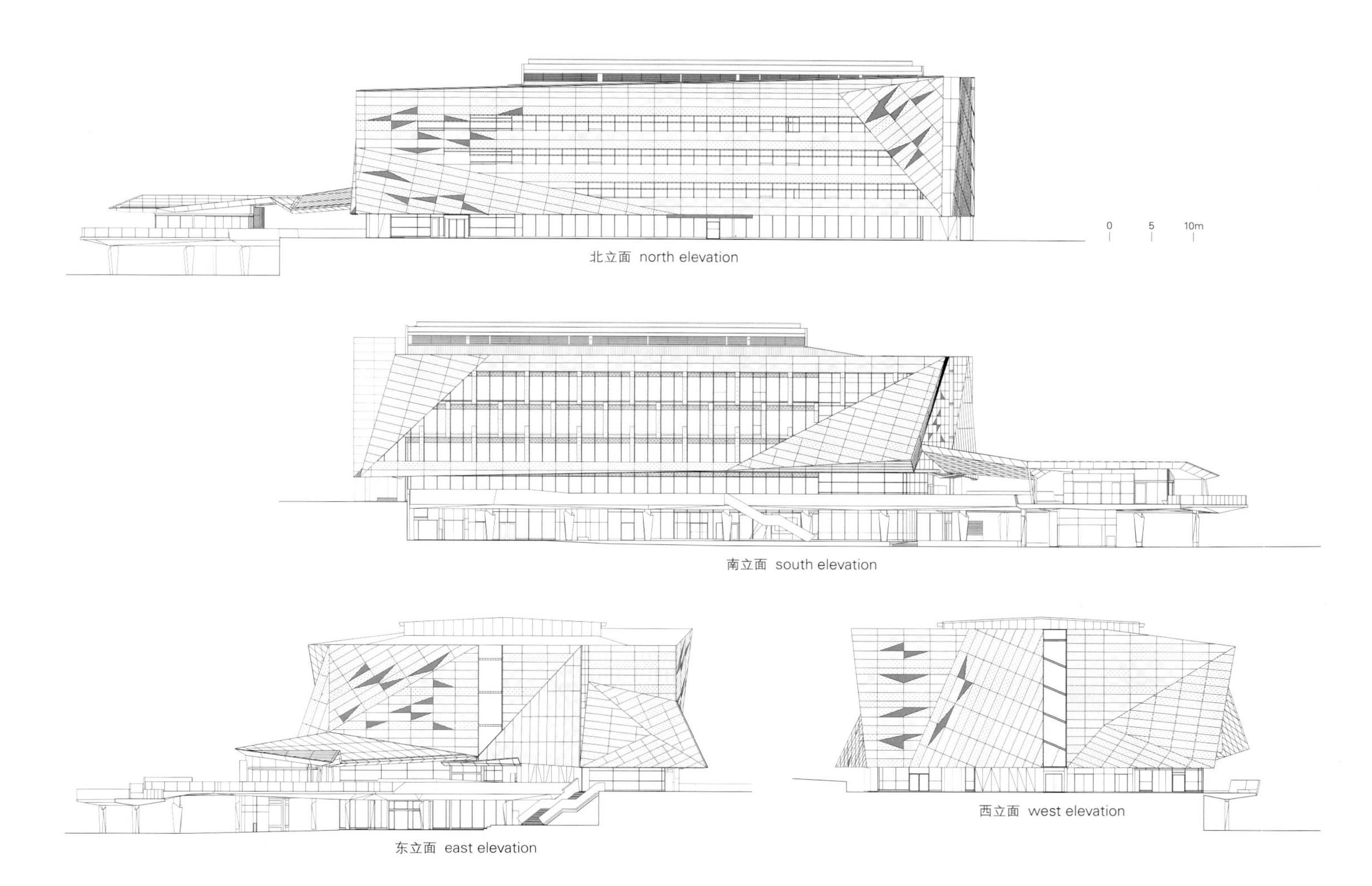

北立面 north elevation

南立面 south elevation

东立面 east elevation

西立面 west elevation

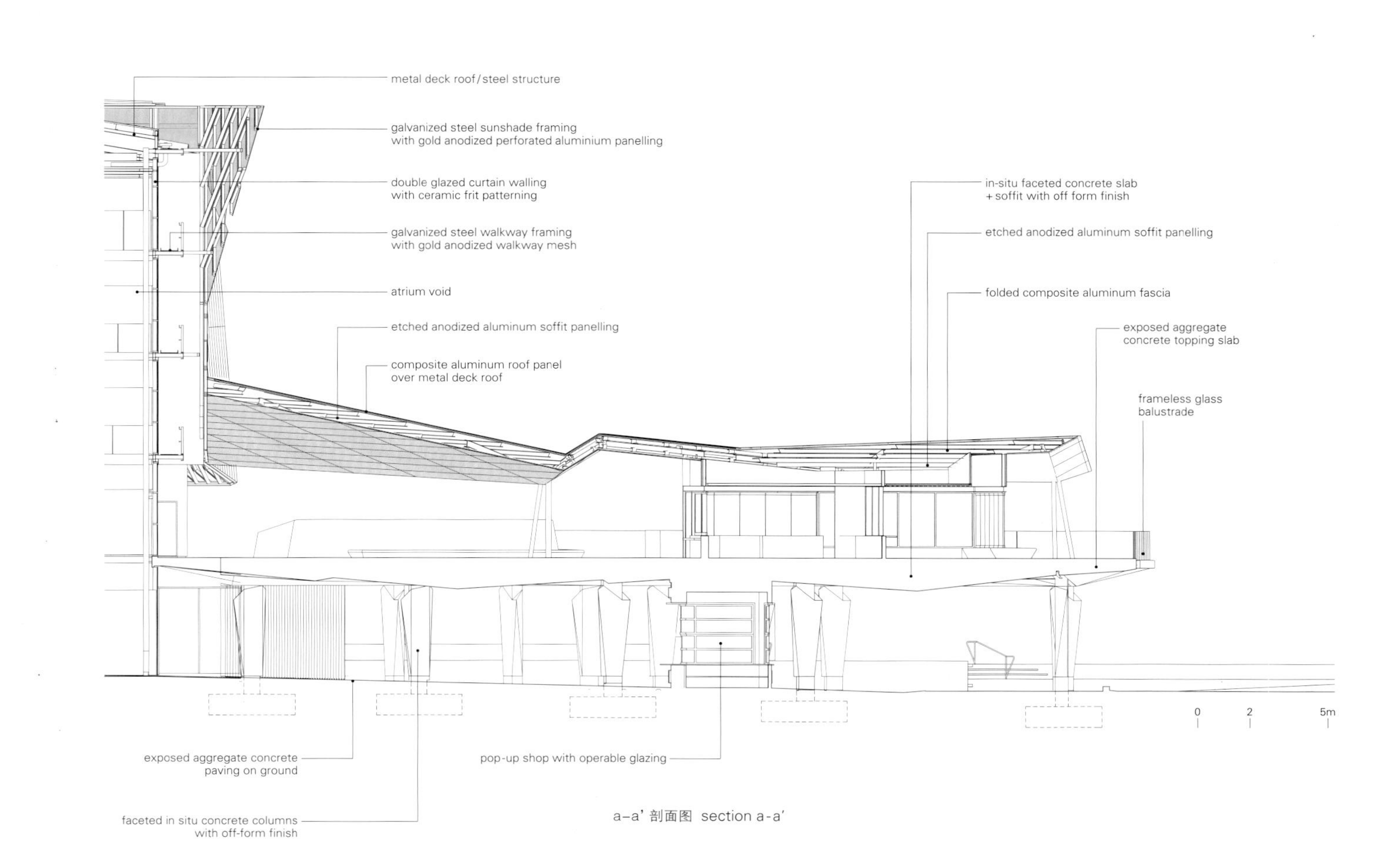

a–a' 剖面图 section a-a'

The focus for the whole building is to create an exciting and high quality new typology of accommodation with an iconic identity and an open, connected and collegiate atmosphere. The site's inherent level changes provide exciting opportunities to create formal and informal landscaped environments linking together to create a "campus street".

The rich architectural palette comprises a wonderful faceted concrete podium and a gold perforated aluminum sun-shading skin, folded and patterned with the feathers of the Carnaby Cockatoo.

The podium, the lower forum, the external skin and the atrium work together to create a building about landscape, which is natural and cultural with the themes: Ngoolark.

courtesy of the architect

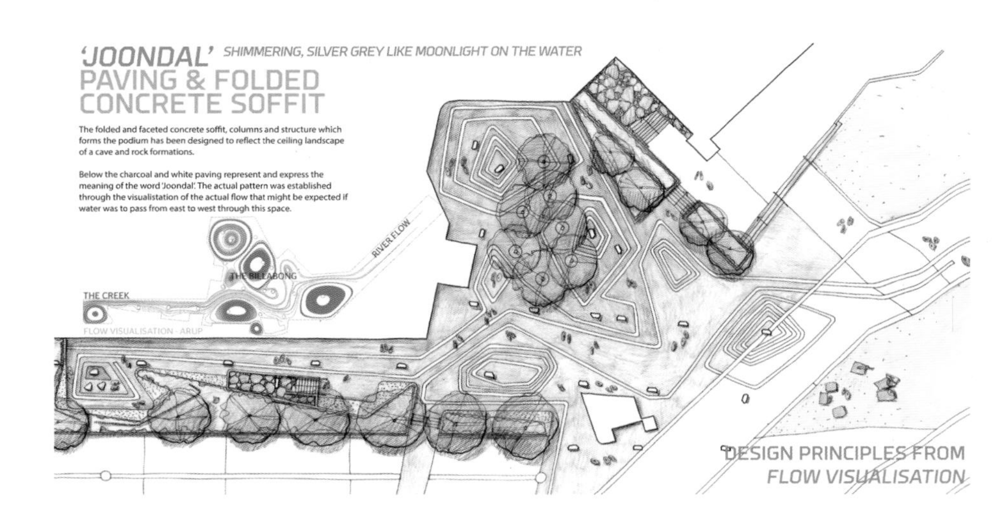

courtesy of the architect

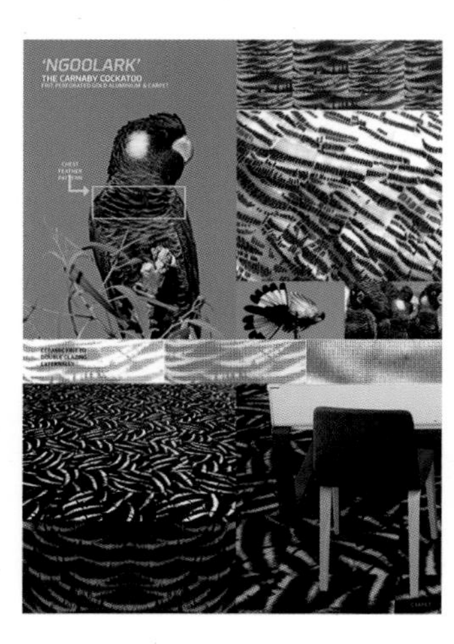

courtesy of the architect

courtesy of the architect

courtesy of the architect

>>136

Brown Meneses Arquitectos

Was founded by two architects Christian Brown and Paola Meneses. Christian Brown was born in 1976 in Ecuador. He studied architecture in Catholic University of Ecuador from 1998 to 2003. Paola Meneses was born in 1978. She also studied architecture in Catholic University of Ecuador from 1998 to 2000 and finished a practical course of Solar Technology Applied to Architecture at the University of Buenos Aires in 2004.

Isabel Potworowski

Has graduated from TU Delft with a Master in Architecture, and currently works for Barcode Architects in Rotterdam. During her graduate studies, she was a member of the editorial committee and wrote several articles for the independent student journal "Pantheon".
Originally from Canada, She completed her Bachelor in Architecture at McGill University in Montreal, where she was awarded the Louis Robertson book prize for the highest grade in first year history. She has also studied for one semester at the Politecnico di Milano. In addition, she has worked at ONPA Architects and Manasc Isaac Architects, both in Edmonton, Canada.

Douglas Murphy

Studied architecture at the Glasgow School of Art and the Royal College of Art, completing his studies in 2008. As a critic and historian, he is the author of The Architecture of Failure(Zero Books, 2009), on the legacy of 19th century iron and glass architecture, and the forthcoming Last Futures(Verso, 2015), on dreams of technology and nature in the 1960s and 70s. Is also an architecture correspondent for Icon Magazine, and writes regularly for a wide range of publications on architecture and culture.

©Vincent Ferran

>>164

Ateliers O-S Architectes

Is managed by three partners Vincent Baur, Guillaume Colboc and Gaël Le Nouëne from the left with complementary characters, which often collaborate with personalities from other fields like plastic art or landscape design. Before they develop their projects, the partners have worked for Rem Koolhaas at OMA and Adriaan Geuze for West 8 in the Netherlands. "O-S refers to many wordplay, in particular Open Source and Operating System, concept that synthesizes ideas of openmindedness and production required for developing a project. With this name they stand out the idea of an open team, where dialog and self expression is fundamental.

>>144

Hascher Jehle Architektur

Is based in Berlin and was founded in 1992 by Rainer Hascher and Sebastian Jehle. They thoroughly analyze the requirements of the site, the function of the building and the wishes of the client. With all this in mind they develop individual architectural solutions, always with the goal of creating high quality space at appropriate costs. The focus of their design concept is always the human being. In the creation of their designs they follow the principle of "simple technology" as little technology as possible, as much as necessary. In this context the considerate treatment of the natural resources has a high priority, because they believe in a building's sustainability as one of its important values.

>>100

D'HOUNDT+BAJART architectes&associés

"Fidelity to the Imagination" is the motto of this architectural office. They always try to create their own language and also apply these to their new projects whether they are small or large. And their major goal is to make architecture a real construction that is as faithful as possible to the picture that they have developed with the client.

C3, Issue 2015.10
All Rights Reserved. Authorized translation from the Korean-English language edition published by C3 Publishing Co., Seoul.

© 2016大连理工大学出版社
著作权合同登记06-2016年第74号
版权所有・侵权必究

图书在版编目(CIP)数据

灵活的学习空间 : 汉英对照 / 韩国C3出版公社编 ; 孙探春等译. — 大连 : 大连理工大学出版社, 2016.7
(C3建筑立场系列丛书)
书名原文: C3: Learning in Fluid
ISBN 978-7-5685-0439-3

Ⅰ. ①灵… Ⅱ. ①韩… ②孙… Ⅲ. ①教育建筑—建筑设计—汉、英 Ⅳ. ①TU244

中国版本图书馆CIP数据核字(2016)第158228号

出版发行：大连理工大学出版社
（地址：大连市软件园路 80 号　　邮编：116023）
印　　刷：上海锦良印刷厂
幅面尺寸：225mm×300mm
印　　张：11.5
出版时间：2016 年 7 月第 1 版
印刷时间：2016 年 7 月第 1 次印刷
出 版 人：金英伟
统　　筹：房　磊
责任编辑：许建宁
封面设计：王志峰
责任校对：王　伟
书　　号：978-7-5685-0439-3
定　　价：228.00 元

发　行：0411-84708842
传　真：0411-84701466
E-mail：12282980@qq.com
URL：http://www.dutp.cn

C3 建筑立场系列丛书 01：
墙体设计
ISBN：978-7-5611-6353-5
定价：150.00 元

C3 建筑立场系列丛书 02：
新公共空间与私人住宅
ISBN：978-7-5611-6354-2
定价：150.00 元

C3 建筑立场系列丛书 03：
住宅设计
ISBN：978-7-5611-6352-8
定价：150.00 元

C3 建筑立场系列丛书 04：
老年住宅
ISBN：978-7-5611-6569-0
定价：150.00 元

C3 建筑立场系列丛书 05：
小型建筑
ISBN：978-7-5611-6579-9
定价：150.00 元

C3 建筑立场系列丛书 06：
文博建筑
ISBN：978-7-5611-6568-3
定价：150.00 元

C3 建筑立场系列丛书 07：
流动的世界：日本住宅空间设计
ISBN：978-7-5611-6621-5
定价：200.00 元

C3 建筑立场系列丛书 08：
创意运动设施
ISBN：978-7-5611-6636-9
定价：180.00 元

C3 建筑立场系列丛书 09：
墙体与外立面
ISBN：978-7-5611-6641-3
定价：180.00 元

C3 建筑立场系列丛书 10：
空间与场所之间
ISBN：978-7-5611-6650-5
定价：180.00 元

C3 建筑立场系列丛书 11：
文化与公共建筑
ISBN：978-7-5611-6746-5
定价：160.00 元

C3 建筑立场系列丛书 12：
城市扩建的四种手法
ISBN：978-7-5611-6776-2
定价：180.00 元

C3 建筑立场系列丛书 13：
复杂性与装饰风格的回归
ISBN：978-7-5611-6828-8
定价：180.00 元

C3 建筑立场系列丛书 14：
企业形象的建筑表达
ISBN：978-7-5611-6829-5
定价：180.00 元

C3 建筑立场系列丛书 15：
图书馆的变迁
ISBN：978-7-5611-6905-6
定价：180.00 元

C3 建筑立场系列丛书 16：
亲地建筑
ISBN：978-7-5611-6924-7
定价：180.00 元

C3 建筑立场系列丛书 17：
旧厂房的空间蜕变
ISBN：978-7-5611-7093-9
定价：180.00 元

C3 建筑立场系列丛书 18：
混凝土语言
ISBN：978-7-5611-7136-3
定价：228.00 元

C3 建筑立场系列丛书 19：
建筑入景
ISBN：978-7-5611-7306-0
定价：228.00 元

C3 建筑立场系列丛书 20：
新医疗建筑
ISBN：978-7-5611-7328-2
定价：228.00 元

C3 建筑立场系列丛书 21：
内在丰富性建筑
ISBN：978-7-5611-7444-9
定价：228.00 元

C3 建筑立场系列丛书 22：
建筑谱系传承
ISBN：978-7-5611-7461-6
定价：228.00 元

C3 建筑立场系列丛书 23：
伴绿而生的建筑
ISBN：978-7-5611-7548-4
定价：228.00 元

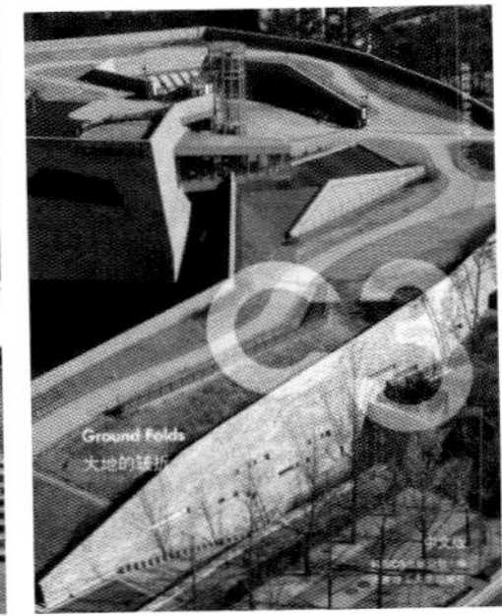

C3 建筑立场系列丛书 24：
大地的皱折
ISBN：978-7-5611-7649-8
定价：228.00 元

C3 建筑立场系列丛书 25：
在城市中转换
ISBN：978-7-5611-7737-2
定价：228.00 元

C3 建筑立场系列丛书 26：
锚固与飞翔——挑出的住居
ISBN：978-7-5611-7759-4
定价：228.00 元

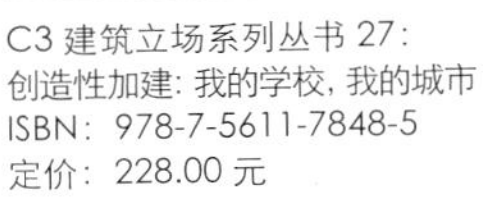

C3 建筑立场系列丛书 27：
创造性加建：我的学校，我的城市
ISBN：978-7-5611-7848-5
定价：228.00 元

C3 建筑立场系列丛书 28：
文化设施：设计三法
ISBN：978-7-5611-7893-5
定价：228.00 元

C3 建筑立场系列丛书 29：
终结的建筑
ISBN：978-7-5611-8032-7
定价：228.00 元

C3 建筑立场系列丛书 30：
博物馆的变迁
ISBN：978-7-5611-8226-0
定价：228.00 元

C3 建筑立场系列丛书 31：
微工作 • 微空间
ISBN：978-7-5611-8255-0
定价：228.00 元

C3 建筑立场系列丛书 32：
居住的流变
ISBN：978-7-5611-8328-1
定价：228.00 元

C3 建筑立场系列丛书 33：
本土现代化
ISBN：978-7-5611-8380-9
定价：228.00 元

C3 建筑立场系列丛书 34：
气候与环境
ISBN：978-7-5611-8501-8
定价：228.00 元

C3 建筑立场系列丛书 35：
能源与绿色
ISBN：978-7-5611-8911-5
定价：228.00 元

C3 建筑立场系列丛书 36：
体验与感受：艺术画廊与剧院
ISBN：978-7-5611-8914-6
定价：228.00 元

C3 建筑立场系列丛书 37：
记忆的住居
ISBN：978-7-5611-9027-2
定价：228.00 元

C3 建筑立场系列丛书 38：
场地、美学和纪念性建筑
ISBN：978-7-5611-9095-1
定价：228.00 元

C3 建筑立场系列丛书 39：
殡仪类建筑：在返璞和升华之间
ISBN：978-7-5611-9110-1
定价：228.00 元

C3 建筑立场系列丛书 40：
苏醒的儿童空间
ISBN：978-7-5611-9182-8
定价：228.00 元

C3 建筑立场系列丛书 41：
都市与社区
ISBN：978-7-5611-9365-5
定价：228.00 元

C3 建筑立场系列丛书 42：
木建筑再生
ISBN：978-7-5611-9366-2
定价：228.00 元

C3 建筑立场系列丛书 43：
休闲小筑
ISBN：978-7-5611-9452-2
定价：228.00 元

C3 建筑立场系列丛书 44：
节能与可持续性
ISBN：978-7-5611-9542-0
定价：228.00 元

C3 建筑立场系列丛书 45：
建筑的文化意象
ISBN：978-7-5611-9576-5
定价：228.00 元

C3 建筑立场系列丛书 46：
重塑建筑的地域性
ISBN：978-7-5611-9638-0
定价：228.00 元

C3 建筑立场系列丛书 47：
传统与现代
ISBN：978-7-5611-9723-3
定价：228.00 元

C3 建筑立场系列丛书 48：
博物馆：空间体验
ISBN：978-7-5611-9737-0
定价：228.00 元

C3 建筑立场系列丛书 49：
社区建筑
ISBN：978-7-5611-9793-6
定价：228.00 元

C3 建筑立场系列丛书 50：
林间小筑
ISBN：978-7-5611-9811-7
定价：228.00 元

C3 建筑立场系列丛书 51：
景观与建筑
ISBN：978-7-5611-9884-1
定价：228.00 元

C3 建筑立场系列丛书 52：
地域文脉与大学建筑
ISBN：978-7-5611-9885-8
定价：228.00 元

C3 建筑立场系列丛书 53：
办公室景观
ISBN：978-7-5685-0134-7
定价：228.00 元

C3 建筑立场系列丛书 54：
城市复兴中的生活设施
ISBN：978-7-5685-0340-2
定价：228.00 元

上架建议：建筑设计

ISBN 978-7-5685-0439-3

定价：228.00元

出版社淘宝店

韩国《C3》杂志中文版已由大连理工大学出版社出版，
欢迎订购！
◆编辑部咨询电话：许老师/0411—84708405
◆发行部订购电话：王老师/0411—84708943

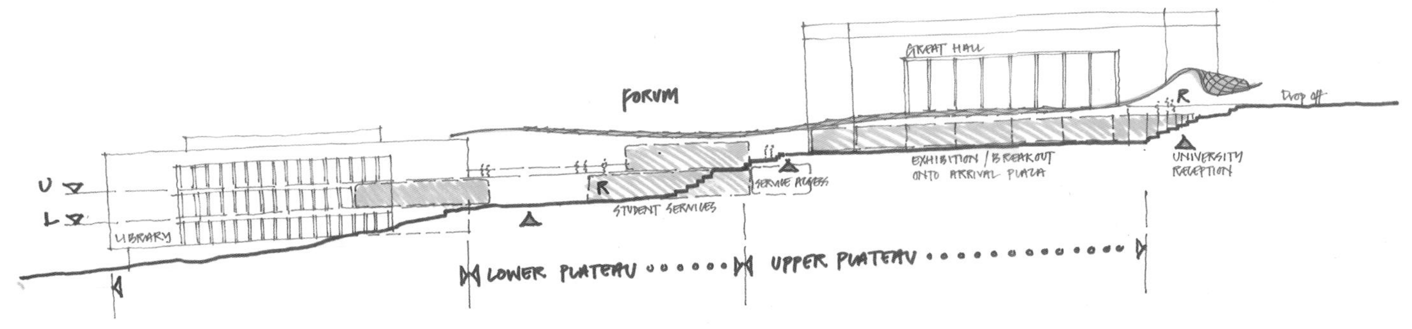
GREAT HALL
FORUM
Drop off
R
EXHIBITION / BREAKOUT
ONTO ARRIVAL PLAZA
UNIVERSITY
RECEPTION
SERVICE ACCESS
R
U
L
STUDENT SERVICES
LIBRARY
LOWER PLATEAU
UPPER PLATEAU

该项目为埃克塞特大学的斯特里汉姆学院校区提供了新的入口，其高耸的由四氟乙烯材料和铜板覆盖的木框架屋顶下容纳了全新的学生设施和教学空间。

论坛将所有必需的设施连为一体，大礼堂、阶梯剧院和学生会办公室等都沿着柱廊式室内商业街设置。商业街内设有咖啡厅、一家商店和一家银行。学生服务中心占据这栋楼的显著位置，拓宽了大学为学生所提供的服务范围，使学生可以获得更好的大学社区服务。另外，论坛集新建的、拥有400个座位的技术一流的礼堂、专门委托设计的公共艺术区、景观式开放空间和整修一新的图书馆于一身，使之成为埃克塞特大学的社交和学术生活中心。

论坛项目的设计起点是斯特里汉姆学院校区的自然特征。埃克塞特大学的斯特里汉姆学院校区以连绵起伏的丘陵地势而闻名。该建筑以贯穿整栋建筑的主要人行路线——"绿色走廊"为主线，其方位和布局以及相邻的景观式广场都与周围的山地景色相得益彰。

该建筑的屋顶结构为网格状外壳，其设计理念主要围绕捕捉光线通过屋顶进入大楼时的光影交错感而形成。通过一种算法来模仿自然的形式，错综复杂的品牌表达模式由此形成。

入口广场的西北端是一座新的入口建筑，作为埃克塞特大学和大礼堂的接待区。这一入口建筑的流畅曲线上升一层的高度，穿过覆顶通道上方的大礼堂的北立面，最后与论坛的最核心部分，即波浪形网格状屋顶，融为一体。

论坛建筑的面积为3500m²，宽敞通风，无柱支撑，拥有同类设计中最大的木屋顶之一。其流畅的外形与原来校园里板板正正的砖结构建筑形成鲜明对照，既与山丘背景浑然天成，又保护了从城市到达特穆尔高原的主要风景。屋顶铜覆层下是独创的混合木结构。在未来的15年中，铜将氧化，形成优雅的铜绿锈表层。仿效传统的全钢结构建造的网格状外壳屋顶，其一系列木构件利用钢节点连接在一起，且在校园里原有建筑之间穿梭。之所以选择木材作为主要建材是因为研究发现在埃克塞特大学建立之前，论坛所在的位置曾是一座植物园。

屋顶使用的是云杉木，而放入窑中烘干的橡木用于室内装饰。屋顶边缘弯弯曲曲的弧线由易于造型的、生长年限较短的橡木装饰。屋顶三角形的网格单元或采用铜表层覆盖，里层是具有隔音效果的橡木条，或是四氟乙烯材料，使整个论坛宽敞的室内空间充满了自然光线。讲台正上方的屋顶天窗则使用传统的小型玻璃嵌板镶嵌，这样可以减小下雨时可能产生的噪音。

玻璃立面将原有建筑物之间的空间围合起来，论坛便位于这片屋顶之下，这是一处通风良好的柱廊式大厅，容纳了一系列全新的以学生为中心的空间。上面一层可以通向大礼堂、学生会办公室和专门为哈佛式探索课程设计的一系列新的学习实验室和研讨室。

埃克塞特大学发现环境因素是影响越来越多本科生选择大学的关键因素，所以论坛的设计目标是达到英国建筑研究院环境评估方法的"优秀"级别。其木结构和铜覆层所体现的可持续性显而易见。除此之外，主建筑物使用过程中自然通风系统、错综复杂的管道制冷技术和节能照明系统等其他特点将会使其对环境友好的表现发挥得淋漓尽致。该建筑物依据SUDS(可持续排水系统)原则而设计，可收集雨水，回收和利用废水，并在校区南部建造新的池塘和湿地。该新项目全部选用低用水的设施。

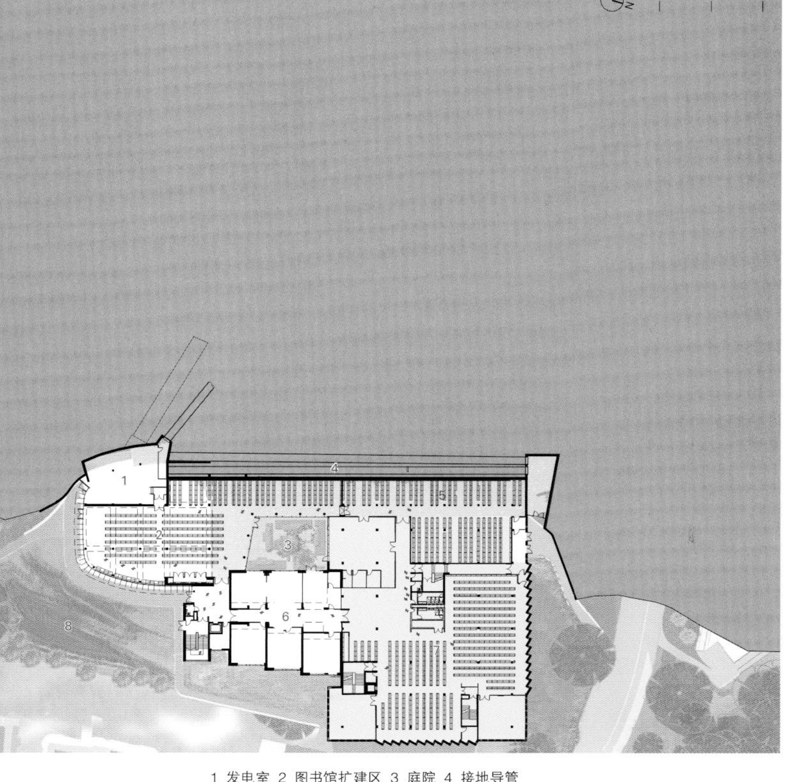

1 发电室 2 图书馆扩建区 3 庭院 4 接地导管
5 可旋转的书架 6 讲室 7 翻新的图书馆 8 洼地
1. plant room 2. library extension 3. courtyard 4. earth tubes
5. rolling book stacks 6. seminar rooms 7. refurbished library 8. swale
低层 lower ground floor

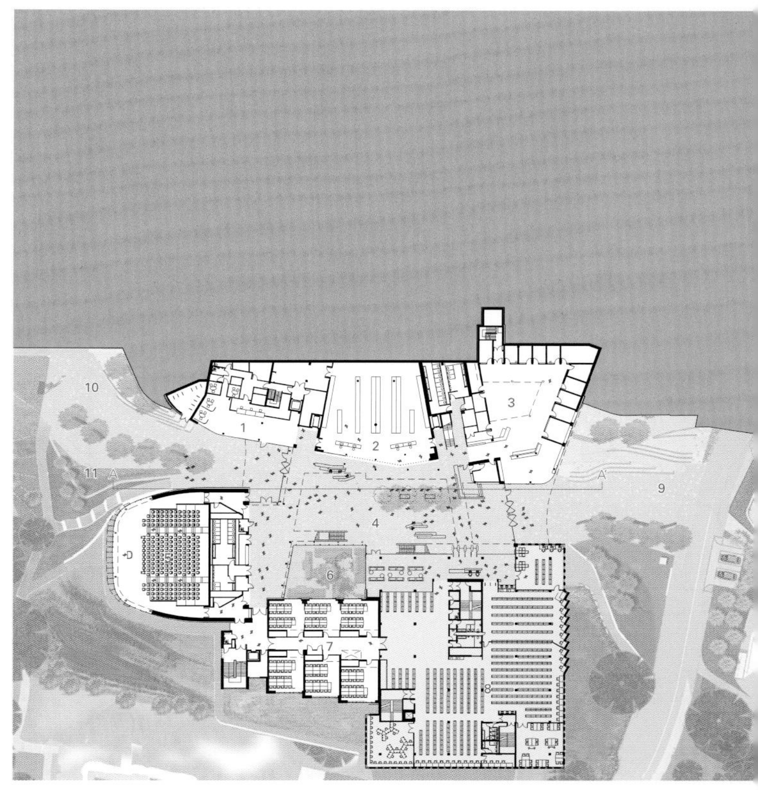

1 银行 2 零售区 3 学生服务中心 4 室内街道 5 礼堂 6 庭院 7 讲室 8 翻新的图书馆
9 北侧的广场 10 南侧的广场 11池塘
1. bank 2. retail unit 3. student services center 4. internal street 5. auditorium 6. courtyard
7. seminar rooms 8. refurbished library 9. north plaza 10. south plaza 11. pond
一层 ground floor

The Forum at University of Exeter

The project provides a new entrance for the University's Streatham campus and houses new student facilities and teaching accommodation beneath a soaring timber-framed ETFE and copper roof.

The Forum links essential facilities such as the Great Hall, lecture theaters and the Student Guild along a galleried, indoor high-street, lined with cafes, a shop and a bank. A student services center takes a prominent position within the new building, broadening and enhancing the range and availability of services that the University offers. A newly built state of the art 400-seat auditorium, specially commissioned public art area, landscaped open spaces and refurbished library complete the Forum, making it the heart of the social and academic life of the University.

Starting point for the Forum project was the natural features of Exeter's famously hilly Streatham campus. The orientation and arrangement of the building and its adjacent landscaped plazas respond to the contours of the hillside setting, which are traced by a "green corridor", the main pedestrian route through the scheme.

Design concept centers around capturing the play of light and shadow as it entered the building through the grid-shell roof structure and they created an intricate brand expression pattern programmed through an algorithm to mimic natural forms.

At the northwestern end of the Entrance Plaza stands a new entrance building that provides a reception area for the University and Great Hall. The smooth curve of the entrance building rises one floor, tracing the northern elevation of the Great Hall above a covered walkway before merging with the centerpiece of the Forum project: an undulating timber gridshell roof designed.

The Forum's roof encloses 3,500m² of an airy, column-free floor area making it one of the largest timber roofs of its kind. Its flowing forms contrasts with the orthogonal brick architecture of the existing campus, responding to the hillside setting and preserving key views across the city to Dartmoor. Beneath the roof's copper cladding, which will gracefully develop

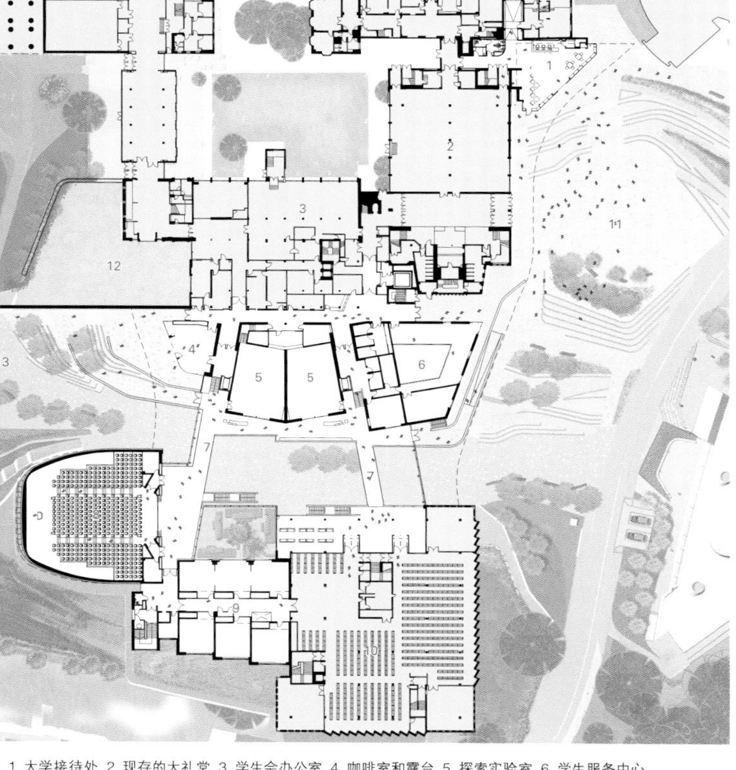

1 大学接待处 2 现存的大礼堂 3 学生会办公室 4 咖啡室和露台 5 探索实验室 6 学生服务中心 7 街道上方的人行桥 8 礼堂 9 讲室 10 翻新的图书馆 11 北侧的广场 12 服务区 13 南侧的广场
1. university reception 2. existing great hall 3. student union 4. cafe and terrace 5. exploration labs 6. student services center 7. bridge over street 8. auditorium 9. seminar rooms 10. refurbished library 11. north plaza 12. service yard 13. south plaza
二层 first floor

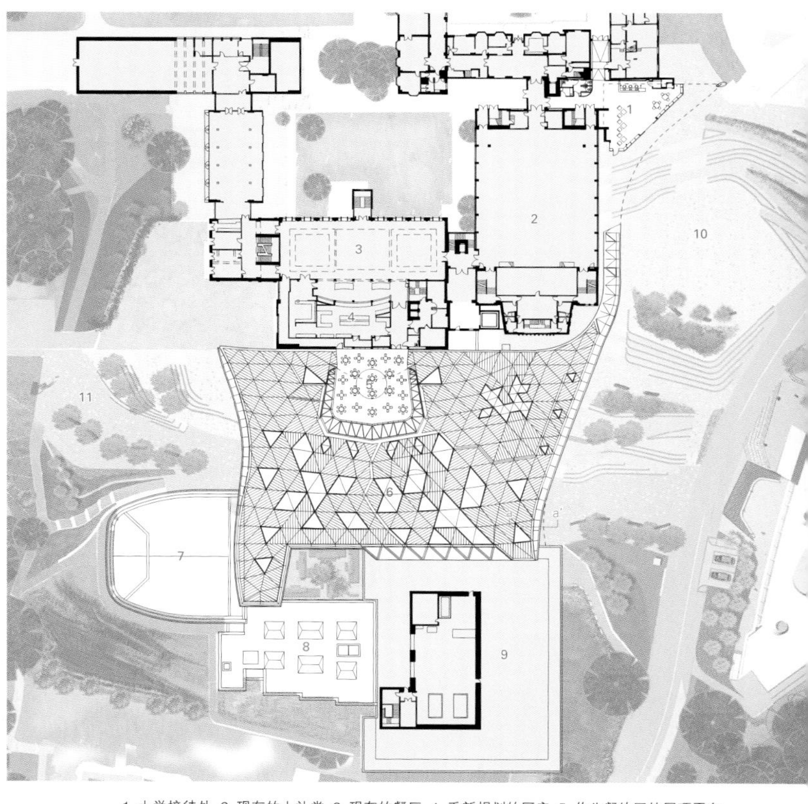

1 大学接待处 2 现存的大礼堂 3 现存的餐厅 4 重新规划的厨房 5 作为餐饮区的屋顶露台 6 网格状屋顶 7 礼堂 8 讲室 9 现存的图书馆 10 北侧的广场 11 南侧的广场
1. university reception 2. existing great hall 3. existing dining hall 4. reconfigured kitchen 5. dining roof terrace 6. grid shell roof 7. auditorium 8. seminar rooms 9. existing library 10. north plaza 11. south plaza
三层 second floor

F O R

项目名称：The Forum, University of Exeter
地点：Streatham Campus, University of Exeter, Devon, United Kingdom
建筑师：Wilkinson Eyre Architects
主要负责人：Stafford Critchlow
项目助理：Chris Donoghue
项目团队：Kevin Bai, Florian Ballan, Chris Davies, Ben Dawson, Paula Friar, Christian Froggatt, Nat Keast, Harsh Lad, Felix Lewis, Leszek Marszalek, Stephen Perrin, Tom Smith, Ivan Subanovic, Simon Vickers, Jan Vogel, Nadine Wagner, Chris Wilkinson, Soo Yau
结构、土木、M&E、消防、音效、AV、IT、交通、通道、BREEAM和环境工程设计：Buro Happold
景观建筑师：Hargreaves Associates
项目经理、质量检测师：Davis Langdon
设计顾问：Turley Associates
甲方：University of Exeter
有效楼层面积：3,500m²
造价：EUR 48 million
竣工时间：2012.4
摄影师：©Hufton and Crow
(courtesy of the architect)

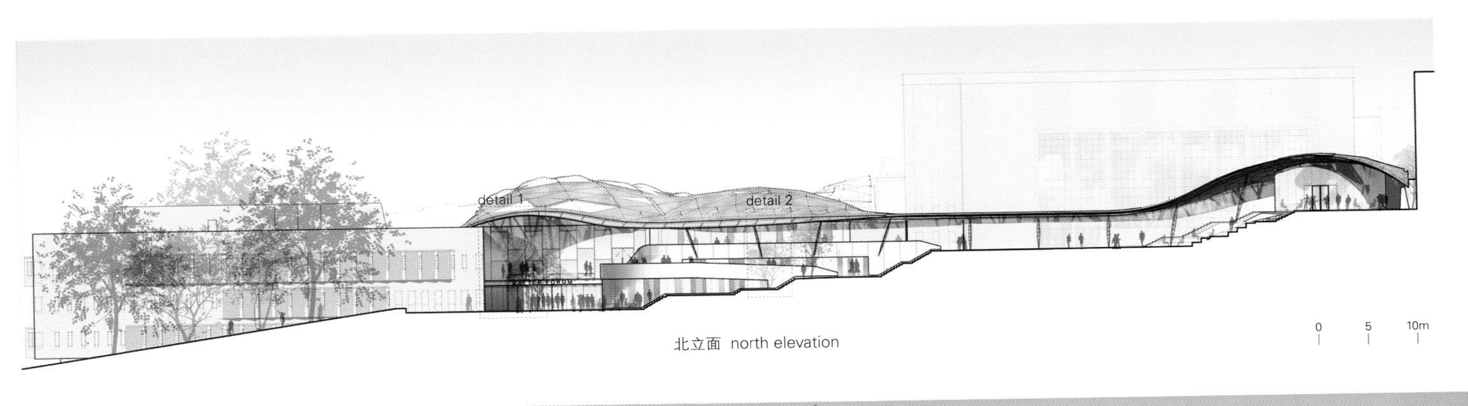

北立面 north elevation

A–A' 剖面图 section A-A'

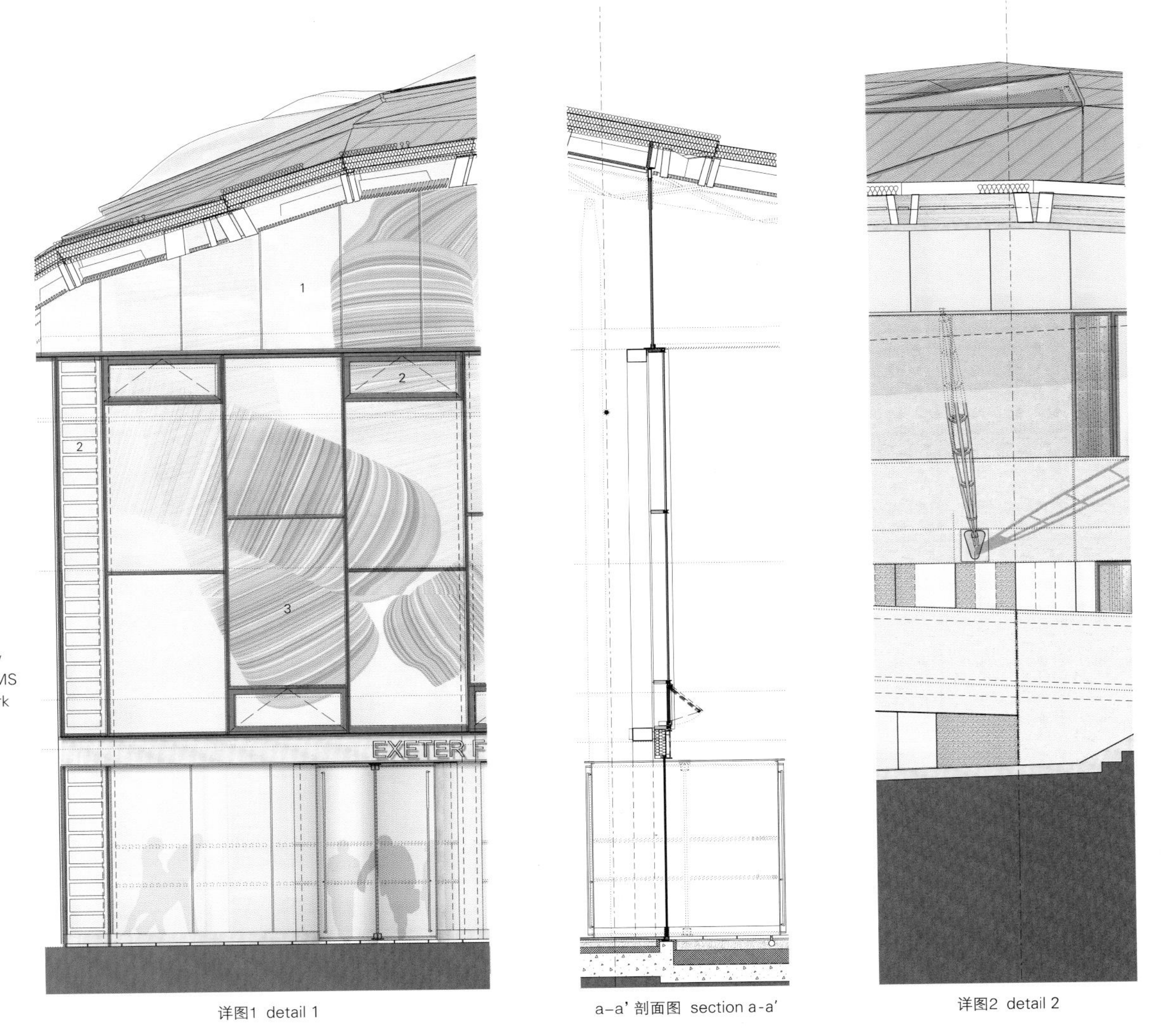

1. shuffle glazed clerestorey
2. opening vent linked to BMS
3. beleschenko glass artwork printed onto both panes of double glazed unit

详图1 detail 1

a–a' 剖面图 section a-a'

详图2 detail 2

COSTA

a verdigris patina over the next 15 years, lies an innovative hybrid timber structure. A matrix of timber members joined at steel nodes emulates the traditionally all-steel construction of a gridshell roof, weaving between the existing buildings of the campus. Wood was chosen as a key material when research revealed that a botanical garden once grew on the Forum site before the University was built.

The structural elements of the roof are Spruce timber, while kiln-dried Oak is used on interior finishes. The sinuous curves of the roof edge are finished in green Oak which adapts readily to the shape. The triangular cells of the grid are either clad in copper, with oak-slatted acoustic baffling on the interior surface, or filled with ETFE, which floods the lofty interior spaces of the Forum with natural light. Smaller panels of conventional glazing are used in the roof windows above teaching areas to mitigate any possible noise caused by falling rain. Beneath the roof and behind the glass facades that enclose the space between the existing buildings, the Forum is an airy, galleried hall which shelters and unifies a series of new student-focused spaces. The upper level gives access to the Great Hall, Student Guild and a new suite of learning labs and seminar spaces designed for Harvard-style exploration sessions.

The Forum was designed to achieve a BREEAM "excellent" rating after the University identified that environmental performance has become key differentiator for an increasing number of undergrads selecting their University. While the sustainability of the timber structure and copper cladding is apparent, other features such as natural stack ventilation, a labyrinth pipe air-cooling system and energy efficient lighting maximize the environmental performance of the main building in use. The building was designed using SUDS principles, with water harvesting, gray water reticulation and new ponds and wetlands created to the south of the campus to assist in water attenuation. Low water-use fixtures were selected throughout the new project.

Ngoolark埃迪斯科文大学学生服务大楼

JCY Architects and Urban Designers

"Ngoolark"（以Noongar语的卡纳比凤头鹦鹉的名字而命名），是埃迪斯科文大学的一栋建筑，它为"大学生活"提供了一个富有活力的平台，是一座互动的、综合的校园建筑，将多样化的功能（为校园社区提供高质量的学生服务）与未来总体规划融为一体。

该项目包括开发一栋全新的校园建筑，通过创建一个集市场、墩墙和集会的公共场所于一身的大学校园，来满足创建埃迪斯科文大学城市生活的需求。

大楼的一层和二层是极具活力的学生枢纽，而其上部楼层的空间结构却较为"私密"，是灵活而优质的办公、研究和创新场所。

整座建筑的设计焦点是建造一种令人兴奋的、高质量的新型建筑，带有标志性的身份特征，并且展现开放的、联系紧密的大学氛围。

项目场地的地势高低不平，为建造正式和非正式的景观环境提供了绝好的机会。这些景观环境连为一体，形成"一条校园街"。

拥有丰富元素的建筑由壮观的、多面的混凝土墩墙和金色穿孔铝质遮阳表皮组成。表皮经过折叠，其图案如同卡纳比凤头鹦鹉的羽毛。

墩座墙、低层的论坛空间、外表皮和中庭共同形成了这座建筑（同时也是一处景观），它具有Ngoolark的自然和文化内涵。

路线规划
course training

路线连接
routes links

城市肌理
urban fabric

费尔德基希Montforthaus文化中心

Hascher Jehle Architektur

新Montforthaus是一个多功能的文化中心。它功能多样，可以举办会议、舞会、展销会、古典音乐会、流行音乐会和戏剧表演。

该建筑和谐地融入费尔德基希这座中世纪古城的历史遗址之中。其相互交织的形式非常现代，而其材质则使用了当地传统的Jura大理石。两者既形成鲜明对照又令人耳目一新，使这一新建筑融入到城镇的现有肌理中。

新文化中心坐落于三个相邻广场融合而成的大型城市空间内，其外观如同当地小镇的河床卵石。空间的自然流动渗透至建筑内部，引导游客进入四层通高的景观门厅，而开放的画廊置身于自然照明的玻璃中庭屋顶之下。

建筑正立面是通透的，邀请过路者进入并直接引导人们进入一个近15m高的明亮的开放式休息室。外部空间无缝渗透到内部空间，门厅完全通透的玻璃墙使周边环境成为空间的一部分，并营造了内部空间的开敞感。一个宏伟且富有雕塑感的楼梯引导访客从入口进入画廊楼层、小礼堂、研讨室，并从那里再上到屋顶露台。

这一新公共建筑总占地面积为12 700m^2。建筑物的核心功能元素直接围绕门厅而分布，可根据需要连为一体或各自独立。敞亮的门厅区域让进入大音乐厅的入口变得宽敞。大小音乐厅与多功能厅相互连接一起形成了单一的大型有机整体，人们从外面也能看到这种造型。

在Montforthaus中心创建声乐环境的目的是为了给音乐家和听众提供更好的条件：音乐家可以在乐队演奏音乐，而观众有可以与在音乐厅相媲美的听觉体验。这非常具有挑战性。有300多平方米的内部表面可以调节移动，比如，天花板上有六处可移动的声帆。在各种各样的设施中，有一些是人工控制的，有一些是自动化的。根据大厅的使用情况，一些特定的声学配置可以激活或停用。这种做法可以使混响时间最优化以满足一些特殊的需求。

与声学系统一样，规划师必须在音乐厅、会议厅、舞厅等不同的功能布置需求之间找到一个平衡点。这些不同的活动都会涉及不同的技术要求，有时候这些不同的技术要求是相互冲突的。因此，某些建筑元素的功能从一开始设计时就是多功能的。

Montforthaus文化中心是按照"绿色和智能建筑"的理念设计的，符合能源效率A级标准。在建筑设计过程中的各个阶段，建筑师都研究调查利用建筑、技术服务、建筑物物理现象之间的协同效应的机会。

INTERSPAR

Montforthaus in Feldkirch

The new Montforthaus is a multipurpose cultural center. It is versatile enough to host conventions, balls, trade fairs, classical concerts, pop concerts and theater performances.

The Montforthaus is harmoniously embedded in the historical urban grain of the medieval old town of Feldkirch. While its formal articulation is demonstratively modern, its materiality picks up the traditional Jura marble of the region, setting up a dialectical frisson between the two while simultaneously weaving the new insertion into the existing fabric of the town. Like a pebble in the riverbed of the town, the new cultural center sits in the flow of urban space between three adjoining squares which fuse into a single large urban space. The same natural flow of space continues into the building, leading visitors into a four-story landscape of foyers and open galleries beneath a naturally illuminated glazed atrium roof.

With its transparent front the Montforthaus is inviting passers-by into the Montforthaus. It leads directly into the almost 15 meter high, brightly lit open foyer. The seamless flow of space from outside to inside and the fully glazed walls of the foyer make the surroundings part of the space and contribute to

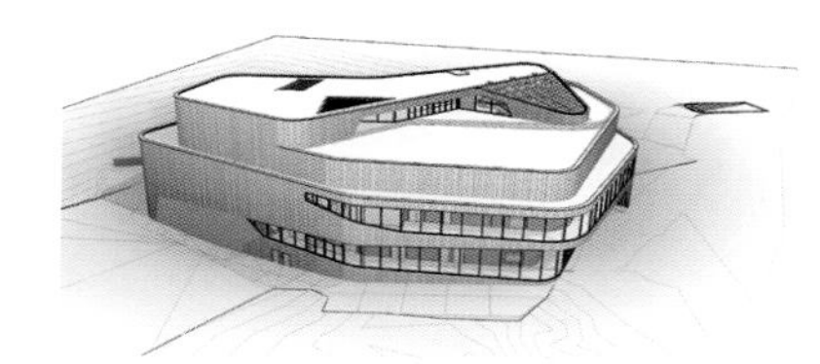

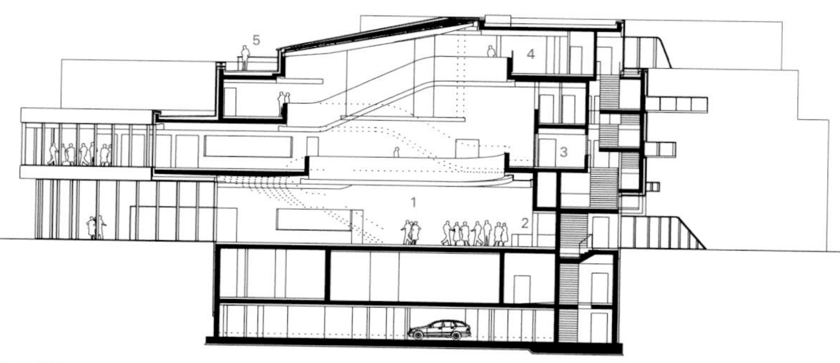

1 门厅 2 衣帽间 3 休息室（讲室） 4 餐厅 5 屋顶露台 1. foyer 2. wardrobe 3. break room (seminar) 4. restaurant 5. roof terrace

A-A' 剖面图 section A-A'

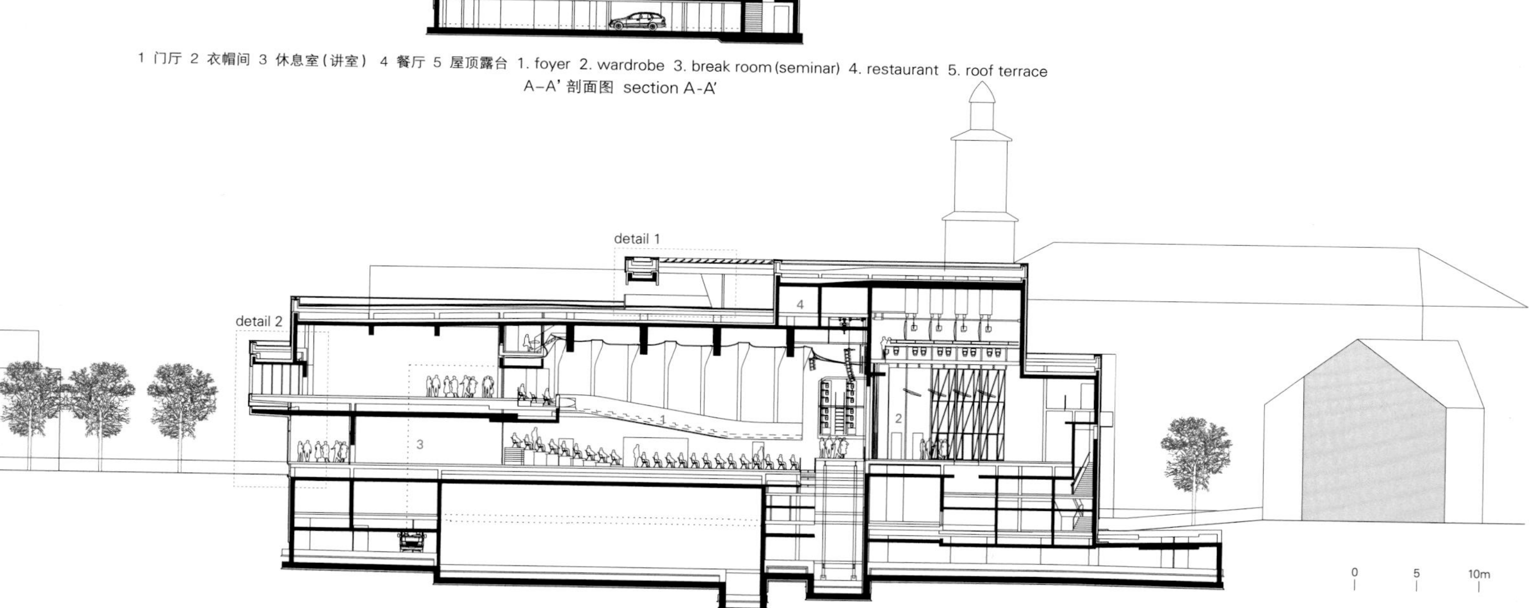

1 大礼堂 2 舞台 3 多功能厅 4 储藏室 1. large auditorium 2. stage 3. multipurpose hall 4. storage

B-B' 剖面图 section B-B'

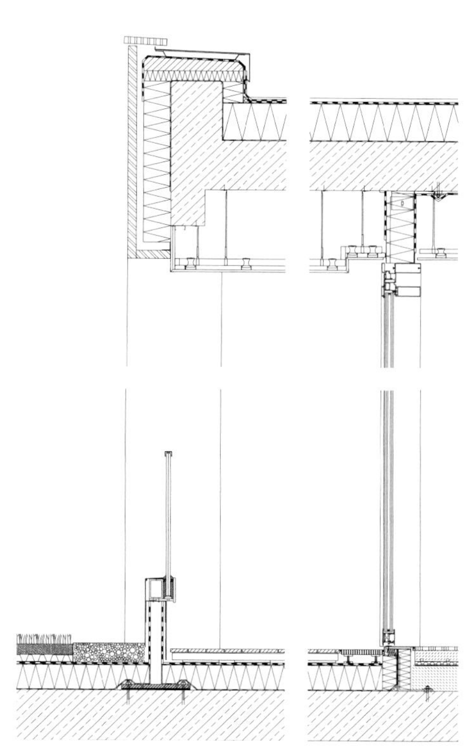

详图1 detail 1

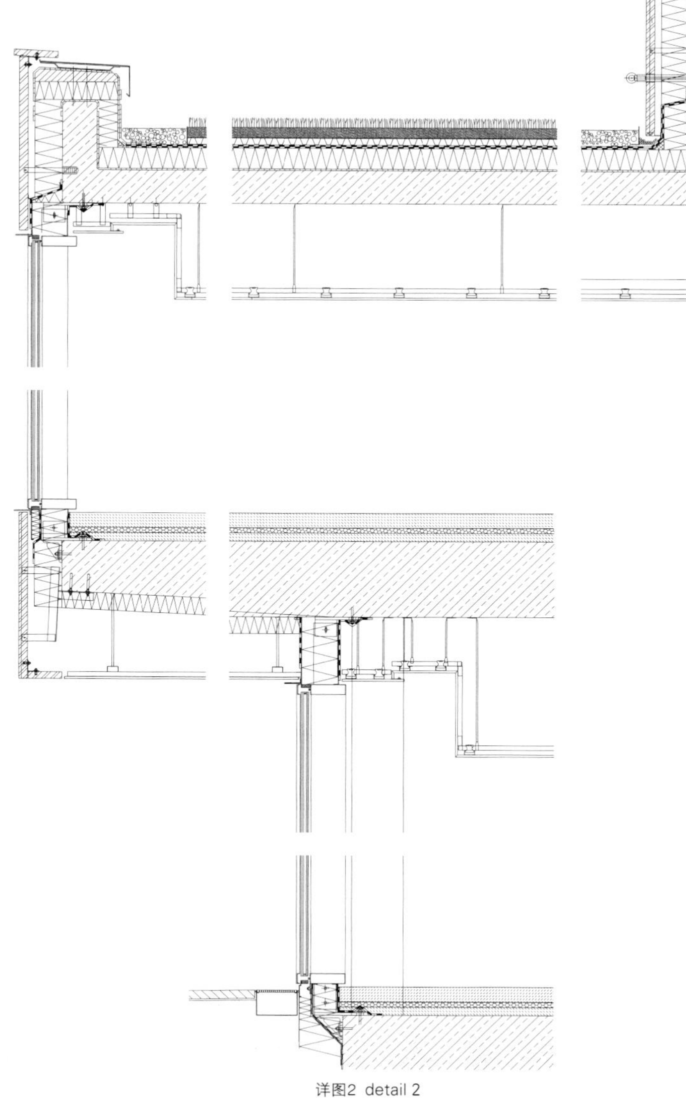

详图2 detail 2

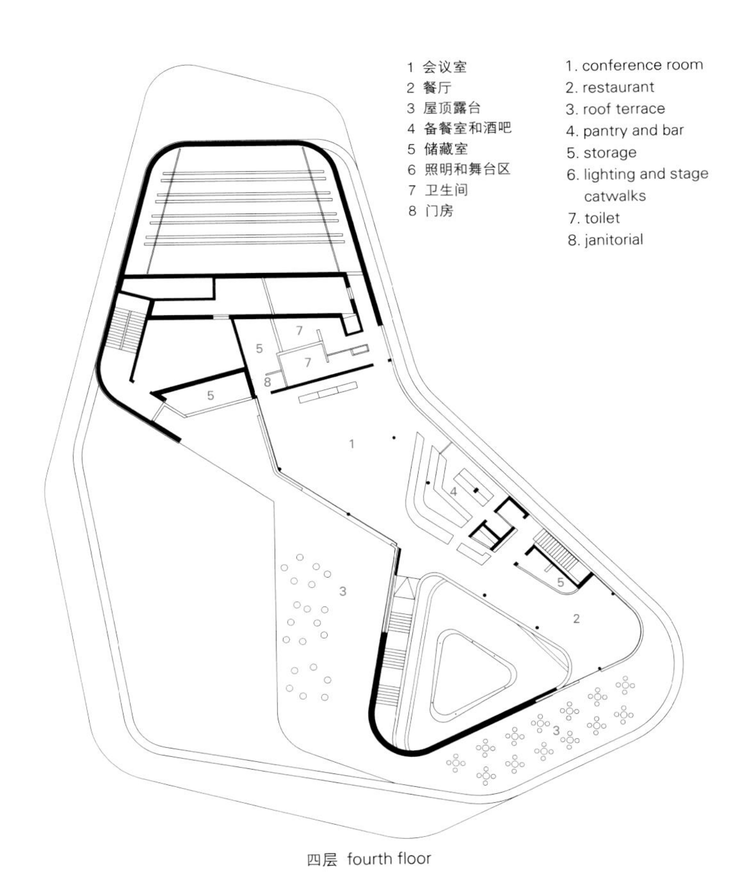

四层 fourth floor

1 大礼堂
2 舞台
3 多功能厅
4 门厅
5 衣帽间
6 备餐室
7 酒吧
8 门房
9 储藏室
10 办公室
11 卫生间
12 会议办公室

1. large auditorium
2. stage
3. multipurpose hall
4. foyer
5. wardrobe
6. pantry
7. bar
8. janitorial
9. storage
10. office
11. toilet
12. conference office

一层 first floor

1 大礼堂
2 大礼堂的阳台
3 小礼堂
4 门厅
5 讲室1
6 讲室2
7 休息室（讲室）
8 备餐室和酒吧
9 档案室
10 储藏室
11 卫生间

1. large auditorium
2. balcony of large auditorium
3. small auditorium
4. foyer
5. seminar room 1
6. seminar room 2
7. break room (seminar)
8. pantry and bar
9. archive
10. storage
11. toilet

二层 second floor

项目名称：Montforthaus in Feldkirch / 地点：Montfortplatz1, A-6800 Feldkirch
建筑师：Planungsgemeinschaft Hascher Jehle Architektur
项目负责人：Frank Jödicke, Gorch Müllauer_Hascher Jehle Architektur, Markus Mitiska_Mitiska·Wäger Architekten
项目团队：Kralyu Chobanov, Christine Dorn, André Flaskamp, Mark Friedrich, Simon Gaier, Lars Gebhardt, Anja Haferkorn, Oliver Heinicke, Wojtek Kaminsky, Carsten Krafft, Beata Maciak, Ralf Mittmann, Ricardo di Parodi, Lubomir Peytchev, Max Porzelt, Dorota Przydrozna, Johannes Raible, Amaya Riero Diaz, Anita Sinanian, Clemence Touzet_Hascher Jehle Architektur, Benjamin Marte, Jürgen Postai, Martin Tschofen, Markus Wäger_Mitiska·Wäger
建筑/施工监督：Baumeister Ing. Michael Hassler / 音效规划：Graner + Partner / 结构工程：Bernard & Brunnsteiner
供暖、通风和电气规划：Dick + Harner, Salzburg / BHM Ingenieure / 建筑物理：IPJ Ingenieure mit ISRW Klapdor / 土壤报告制作：Geotek
测量师：Markowski / 舞台技术规划和照明设计：LDE / 消防安全规划：IBS / 媒体技术规划：Graner+Partner
用地面积：11,689m² 总建筑面积：2,552m² / 有效楼层面积：13,435m² 体积：60,558m³
竞标时间：2008.8 / 设计时间：2011.3 / 施工时间：2012.10~2015.1
摄影师：©Svenja Bockhop(courtesy of the architect) (except as noted)

the sense of an expansive interior. A broad sculptural stair leads visitors up from the entrance to the gallery levels, the small auditorium, seminar rooms and from there onto the roof terrace.
The new public building provides a total of 12,700m² of gross floor area. The core functional elements of the building are distributed directly around the foyer and can be joined or separated as required. The extensive foyer zone forms a spacious entry to the large concert hall. The large and small concert halls as well as the multipurpose spaces are articulated as a single large organic volume, the form of which can be seen from outside.
The aim of creating an acoustic environment in the Montforthaus that offers both the musicians playing in the orchestra and the audience an aural experience comparable to that of a philharmonic hall was a significant challenge. More than 300m² of the internal surfaces have been made adjustable, for example 6 movable acoustic sails in the ceiling. Of the various elements, some are motorized and some are mechanically operated – depending on the use of the hall – so that specific acoustic configurations can be activated or deactivated. As such, the reverberation time can be optimized to meet specific requirements.
As with the acoustics, the planners had to find a balance between the very different staging demands of concerts, conventions, balls, etc. Each of these different kinds of events involves aspects that have different and sometimes conflicting technical requirements. As a consequence, certain building elements were designed to be multi-functional from the outset.
The Montforthaus has been planned in accordance with "Green and Smart Building" concepts and conforms to Energy Efficiency Class A. During all stages of the design process, opportunities to exploit synergy effects between the architecture, technical services and building physics were investigated.

维谢格拉德镇中心

aplusarchitects

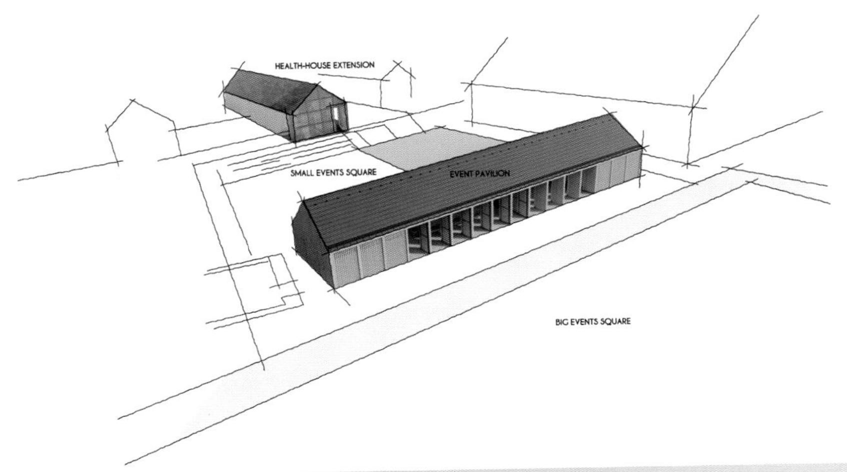
HEALTH-HOUSE EXTENSION
SMALL EVENTS SQUARE
EVENT PAVILION
BIG EVENTS SQUARE

维谢格拉德镇有1800户居民，被认为是匈牙利最小的城镇。镇中心的发展规划向人们很好地展示了经济危机对整个设计规划方案的影响。除了其他的负面影响外，经济危机也导致建筑师不得不修改最初大胆而夸张的设计想法，同时还促使这一可持续发展设计的出现，既可满足当地社区的日常需求，也可满足旅游业的临时需求。

维谢格拉德镇有一千年的历史，这里曾是国王的行宫。然而，无论是从小镇规模还是从其建筑来看，现在的小镇都显得原始而纯朴，充满乡土气息。2000年，维谢格拉德这个地方变成维谢格拉德镇，于是开始着手一项"宏伟"的计划（与小镇规模相比，可谓"宏伟"）。小镇推出一场设计竞赛，在原来曾是中世纪城市主广场的地方修建一个新的城镇中心。aplusarchitects建筑事务所最终成为这场设计竞赛的赢家。在2008年，aplusarchitects建筑事务所开始设计全新的镇中心、镇政厅、公共大厅、文化中心、新的中央广场和其他公共空间。然而，在接下来的几年爆发了经济危机，小镇不得不改变以前的想法，来修改设计方案。考虑到可持续发展和成本效益，镇中心的发展规划有了新的方向。建筑师们完全重新定义了设计方案，设计重点放在对现有的和闲置的建筑物的重新改造利用方面，并且少建新建筑，将重点放在现代化、翻新整修和扩建方面。唯一一个全新元素是活动大厅。

活动大厅位于最初设计的镇政厅的位置，在这个中世纪德国小镇主广场的正面。一方面，活动大厅可以作为镇中心的脸面，成为通向多瑙河的大门；另一方面，活动大厅将大小活动广场连为一体。朝向广场两侧的门都可以打开，形成叶片一样的薄墙。如果一侧的门打开，活动大厅就为这一侧的广场活动提供服务。如果同时在大小广场举行盛大活动，两侧的门都会打开。无论是开放的还是关闭的，活动广场都可以举办展览、演讲展示、招待会，也可以成为工艺品市场。

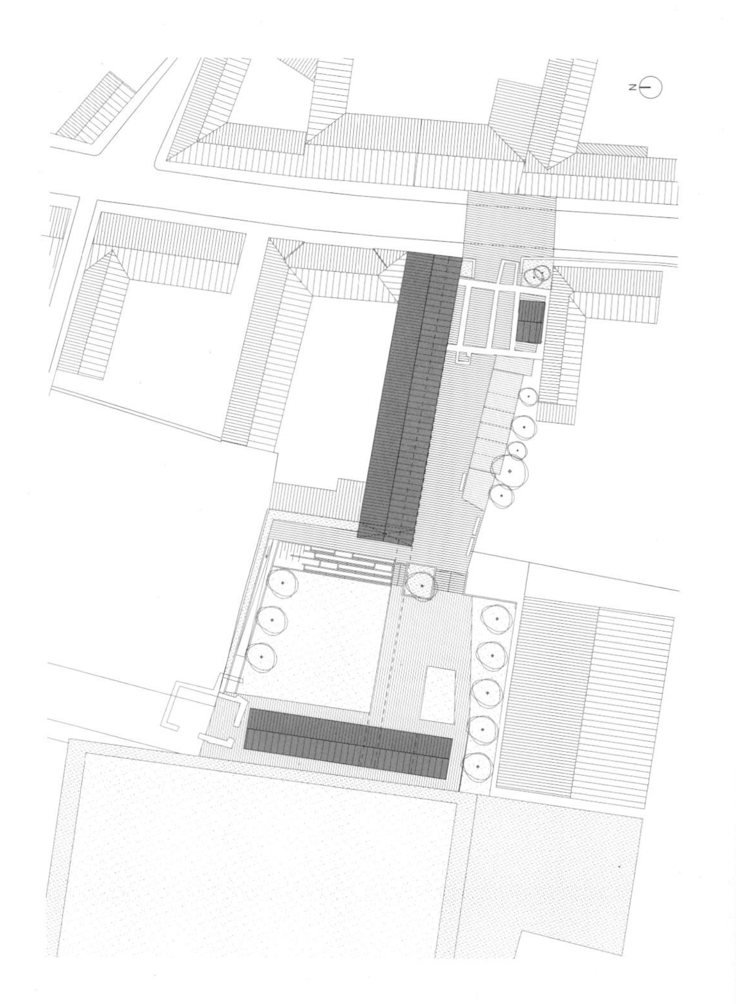

健康之家 (Health-house) 的外立面用嵌板进行扩建，是对小镇主街上可以俯瞰多瑙河那些房屋华丽的木构山形墙的现代诠释。实际上，此处是通往阁楼的户外爬梯，然而这一侧正对着活动广场和多瑙河，因此其主要功能是成为健康之家的外立面。

对现有公共空间的翻新整修主要是为了满足小镇当前的实际需要。作为公共空间设计规划的一部分，建筑师重新改造了健康之家的庭院，周末时将其用作生物市场。活动大厅和健康之家之间这个较小的活动广场，完全根据小镇的规模而设计规划，可供小镇组织小型活动使用；而较大的活动广场，绿草茵茵，朝向多瑙河，在举行举世闻名的"城堡节"期间，这里将充满生机。

Visegrád Town Center

Visegrád, with its 1800 inhabitants, is considered to be the smallest town of Hungary. The development of the town center is a fine example for how the original exaggerating ideas were altered due to the economic crisis – besides its negative effects – and facilitated the birth of a sustainable development, satisfying the continuous needs of the local community and the temporary demands of tourism.

Visegrád is a thousand-year-old settlement, formerly used as royal residency, however the town today is rather rustic in terms of its scale and buildings. In 2000, after acquiring the town status, the settlement undertook a monumental mission, compared to its size. At the place of the supposed medieval town's main square, a plan was put out to competition to build a new town center. In 2008, as the winners of the previously mentioned competition, aplusarchitects began to design the town center, town hall, saloon, cultural center, a new main square and other public spaces. However, in the next few years the emergence of the economic crisis overwrote the original ideas and required the revision of the design program. Considering sustainability and cost-effectiveness the development of the town center took a new direc-

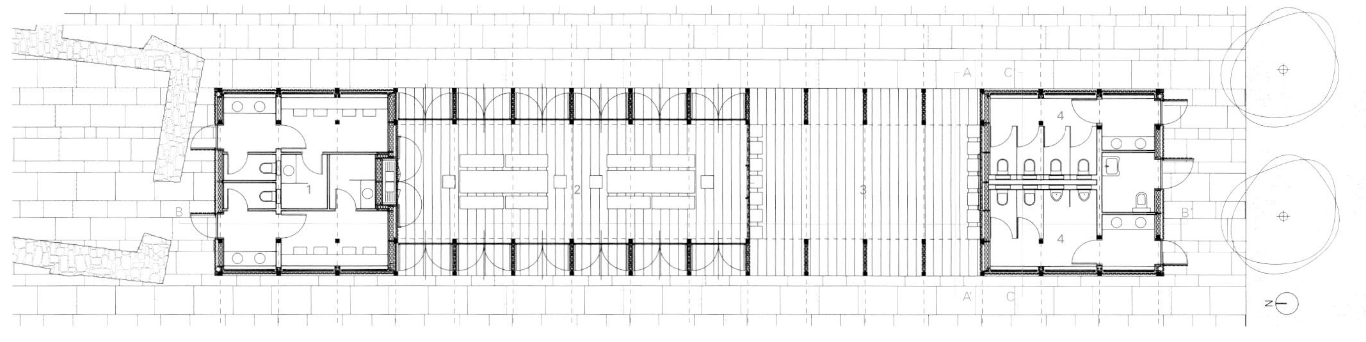

1 更衣室和淋浴间 2 活动室 3 覆顶的开放空间 4 卫生间
1. dressing room and shower 2. event room 3. covered open space 4. toilets
一层 ground floor

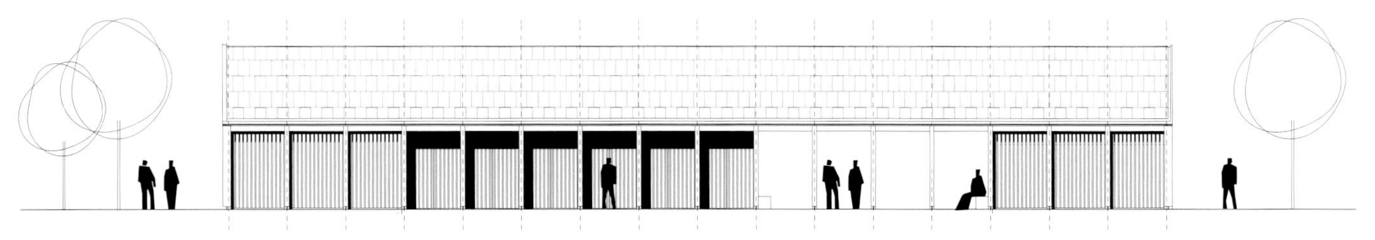

西立面 west elevation

1. roof tiling
2. fly screen
3. tinning RAL 7016
4. bitominous waterproofing
5. zink metal flashing RAL 7016
6. flames and mushroom-free timber 5/15
7. flames and mushroom-free batten 7,5/7,5
8. flexible filling
9. wooden window(4-18-4, Ug=1,1 W/m²K
10. varnished pine pillar 15/15
11. larch boarding
12. mineral wool insulation 15cm
13. air gap
14. steel bracket
15. aluminum clamping profile
16. XPS insulation board
17. unique steel shoe
18. andesite stone slab
19. mineral wool insulation 20cm
20. CD50 metal profile
21. moisture-resistant gypsum board ceilings
22. timber wall frame system 5/15
23. trespa virtuon HPL panel steel grey 8mm
24. trespa wall frame system
25. vapor barrier(Tyvek SD2 VCL)
26. trespa virtuon HPL panel steel grey 8mm
27. trespa TS200 hidden fastening
28. cast floor

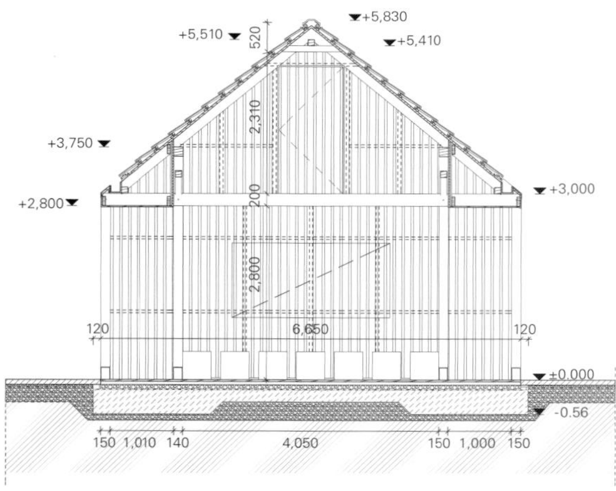

A-A' 剖面图 section A-A'

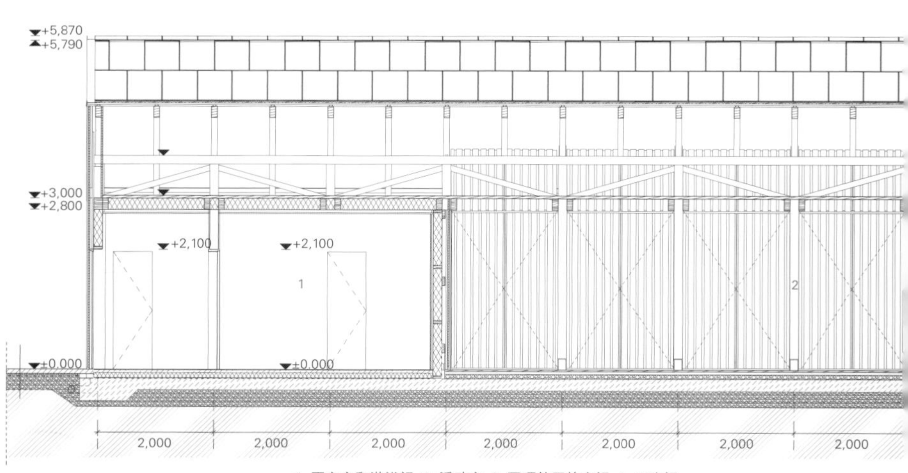

1 更衣室和淋浴间 2 活动室 3 覆顶的开放空间 4 卫生间

1. dressing room and shower 2. event room 3. covered open space 4. toilet

B-B' 剖面图 section B-B'

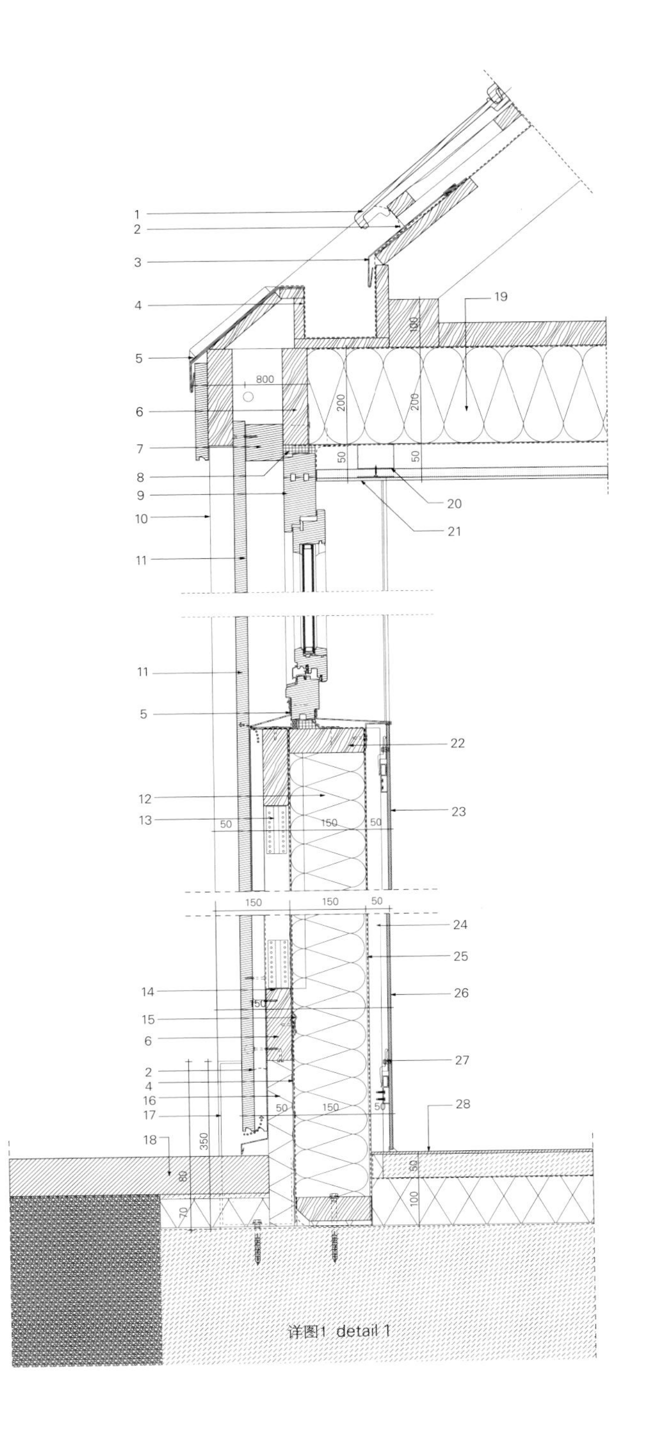

详图1 detail 1

tion. The architects completely redefined the design program, which put the emphasis on the utilization of the existing and idle buildings. Moreover, modernization, refurbishment and expansion were major concerns with a low number of new constructions. The only entirely new element is the events hall.

The events hall can be found at the place of the town hall, designed originally, in the facade of the medieval German town's main square. Its role, on the one hand is to provide a face for the town center with being a gate towards the Danube, on the other hand it creates connection between the big and the small events squares. The building can be opened into blade walls, and with its "disappearing" gates it is able to serve one and the other, in case of grand events, both. Its events square – open or closed – is suitable for organizing exhibitions, presentations, receptions and craft markets.

The paneling expansion of the Health-house's facade is the contemporary transcript of the ornate wooden pediments of the houses in the Main Street overlooking the Danube. Its actual function is an outdoor covered-open loft ladder, however its main role is to form the facade of the Health-house towards the events square and the Danube.

The refurbishment of the public spaces is adjusted to the current and real needs of the town. As part of the public spaces program, the yard of the Health-house was formed, which is used as a bio market at the weekends. The smaller events square between the events buildings and the Health-house is suitable for organizing minor town programs, a public square calibrated to the size of a small town. The bigger grassy space, facing the Danube, comes alive at the time of the internationally-known Palace Games.

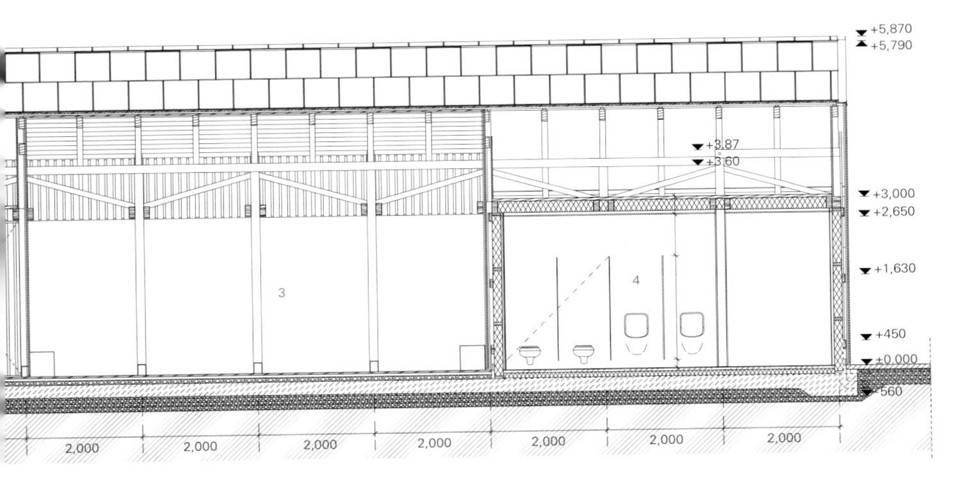

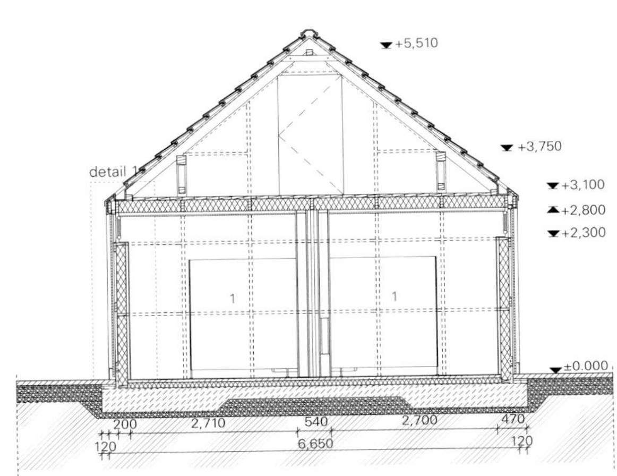

1 卫生间 1. toilets

C–C' 剖面图 section C-C'

项目名称：Visegád Town Center
地点：Visegrád, Hungary
建筑师：aplusarchitects
设计师：Tamás Anna Mária, Kovács-Andor Krisztián
景观建筑师：S73 stúdió
甲方：Municipality of Visegrád
用地面积：7,341m²
总建筑面积：370m²
设计时间：2012
施工时间：2013—2014
竣工时间：2014
摄影师：©Ákos Mátételki and József Lipka (The Greypixel Workshop) (courtesy of the architect)

圣日尔曼阿尔帕容文化中心

Atelier O-S Architectes

在2010年，圣日耳曼阿尔帕容镇就开始要修建一座新的设施，内设媒体图书馆和一所音乐舞蹈学校。原有的音乐舞蹈学校和公共图书馆位于一栋又小功能又不完善的建筑物内，这个文化中心完工后将会取代原来的音乐舞蹈学校和图书馆。

市政府选择的场址呈长而狭窄的带状，西边是尚特卢公园，东边是L' Orge山谷，地势自西向东倾斜，下面便是山谷。另外，文化馆的一边是罗兰·加洛斯高中，另一边是墓地。

这个项目被设计成环形，其顶部广场和底部广场相连。文化中心的主体结构有图书馆、音乐舞蹈学校和一个中央大堂（包括动画空间、礼堂和展览空间等）。整座建筑的功能分布和行政管理区域都是围绕中央大堂垂直分布的。

该建筑一层入口靠近Leuville路一侧，与地面齐平，一面顺着斜坡向下延伸，从Leuville路一侧看非常醒目。因此，建筑的主入口就设在此处，前面是空间非常宽敞的矿石建成的广场。这个文化中心的公共入口位于该建筑两个分支之间，如同一只手，欢迎来访者走进建筑内。

媒体图书馆和大厅位于上层。这里是整座建筑的核心，是来访者查询资料的地方。图书馆呈L形布局，大而宽敞的阅读区域由彩色家具装饰，且面向L' Orge山谷开放，将其景色尽收眼底。在下面一层，音乐

舞蹈学校、行政办公室和工作室位于北侧，并且两个主要房间的视野非常开阔，提供了望向山谷的视野。

技术保障部门和自行车停放处位于公共广场的底层。

在较低的公共广场一层，建筑物的底部被挖空，使上面部分能够悬浮在空中。图书馆主要靠巨大的预应力混凝土横梁来支撑。为了保持整体结构的平衡，预应力混凝土横梁靠埋在地下的一个巨大体量来支撑。

建筑的西侧外立面和东侧外立面分别面向城市和L' Orge山谷。西侧外立面朝向地势较高的广场，而东侧外立面形成山谷视野，其下面是地势较低的广场，一直延伸到远处的天际线。

通过精心设计的透明性，这一新建的公共建筑向人们展示了这座城市和山谷的优美景色，让人们从外面就能看到里面举行的文化活动，更重要的是人们可以将山谷的风景尽收眼底。

外墙板的设计一丝不苟，交替使用的玻璃板和印刷板材的色调明暗不同，使这一公共建筑外立面充满节奏感。

坚固的墙板由精心测量计算的挤压铝元素建造，既营造了一种随机感，同时可以隐藏所有的机械装置。

屋顶由绿色植被覆盖，充分提高了建筑的活力，人们在此可以愉快地欣赏周围的建筑。

这座新的建筑将成为场地的发电机，内有无限的潜能，将当之无愧地成为圣日耳曼莱阿尔帕容市的文化和社会中心。作为该市的组成部分，以及城市景观的延伸，再加上清新的建筑风格，这座公共建筑成为当地一个旅游景点，同时将Leuville路和L' Orge山谷连为一体。

此项目设计方案综合考虑了其地理位置、形状、建筑成本和材料，旨在促进交流，为公众提供一个欢乐、聚会之所。

Saint-Germain-lès-Arpajon Cultural Center

In 2010, the town of Saint-Germain-les-Arpajon wants to acquire new equipment comprising a media library and a school of music and dance. This cultural center will replace the current school of music and dance and the current public library which are located in a small and malfunctioning existing building.
The site chosen by the city is a long and narrow strap plot which fits in a scale between Chanteloup park overlooking the West and L'Orge Valley in the East. The plot is steep and it looks down to the valley. It's framed by Rolland Garros High School on a side and by a cemetery on the other side.
The program is organized as a continuous loop, joining the top square to the bottom square. The structural programmatic entities of the cultural center are the library, the school of music & dance and a centralized lobby(including animation spaces, auditorium and exhibition spaces) from where the whole building and the administration are vertically distributed.
At the Leuville Road, the project is approaching the ground to emerge from it down below the slope. The equipment clearly appears from the road Leuville, thus affirming this main access with a mineral and generous square. The public entrance of the cultural center opens between the two branches of the equipment as a welcoming hand to guide you through the building.
The media library and the lobby are located at higher level. This is the heart of the project, this is where visitors find information. The library is organized as a L-plan on a large open reading spaces structured by colored furniture. The library is widely opened towards the L'Orge Valley. On the lower level, the music & dance school, administrative offices, and lecture studios are located on the north side, while the two main rooms offers generous views to enjoy the landscape of the valley.

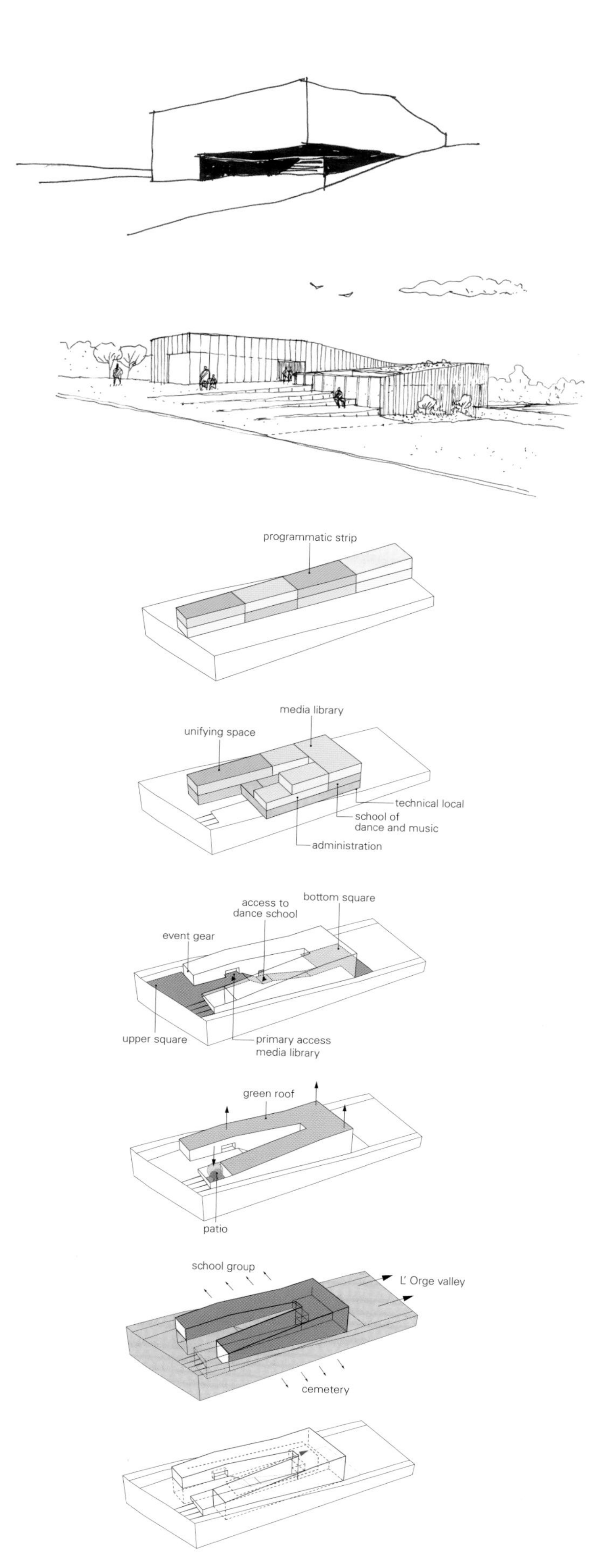

courtesy of the architect

南立面 south elevation
东立面 east elevation
北立面 north elevation
西立面 west elevation
0
5
10m

项目名称：Cultural Center at Saint-Germain-lès-Arpajon / 地点：Saint-Germain-lès-Arpajon, France
建筑师：Ateliers O-S Architectes
项目团队：Vincent Baur, Guillaume Colboc, Gaël Le Nouëne,design phase _ Pierre Teisseire, Jeremie Galvan, construction phase _ Vincent Menuel
景观设计师：OLM / 服务工程师：CFERM / 结构工程师：C&E / 外围护结构设计工程师：ARCORA / 建筑监理：MDETC
音效工程师：ORFEA / 标志设计师：BEAU-VOIR
用地面积：2,848m² / 总建筑面积：2.173m² / 有效楼层面积：2,407m² / 竞赛时间：2011.1 / 施工时间：2013—2015 / 竣工时间：2015.2
摄影师：
©Vincent Baur(courtesy of the architect)-p.167top, p.170, p.172
©Vincent Ferrané(courtesy of the architect)-p.164~165, p.168~169, p.171, p.173, p.174, p.177, p.178

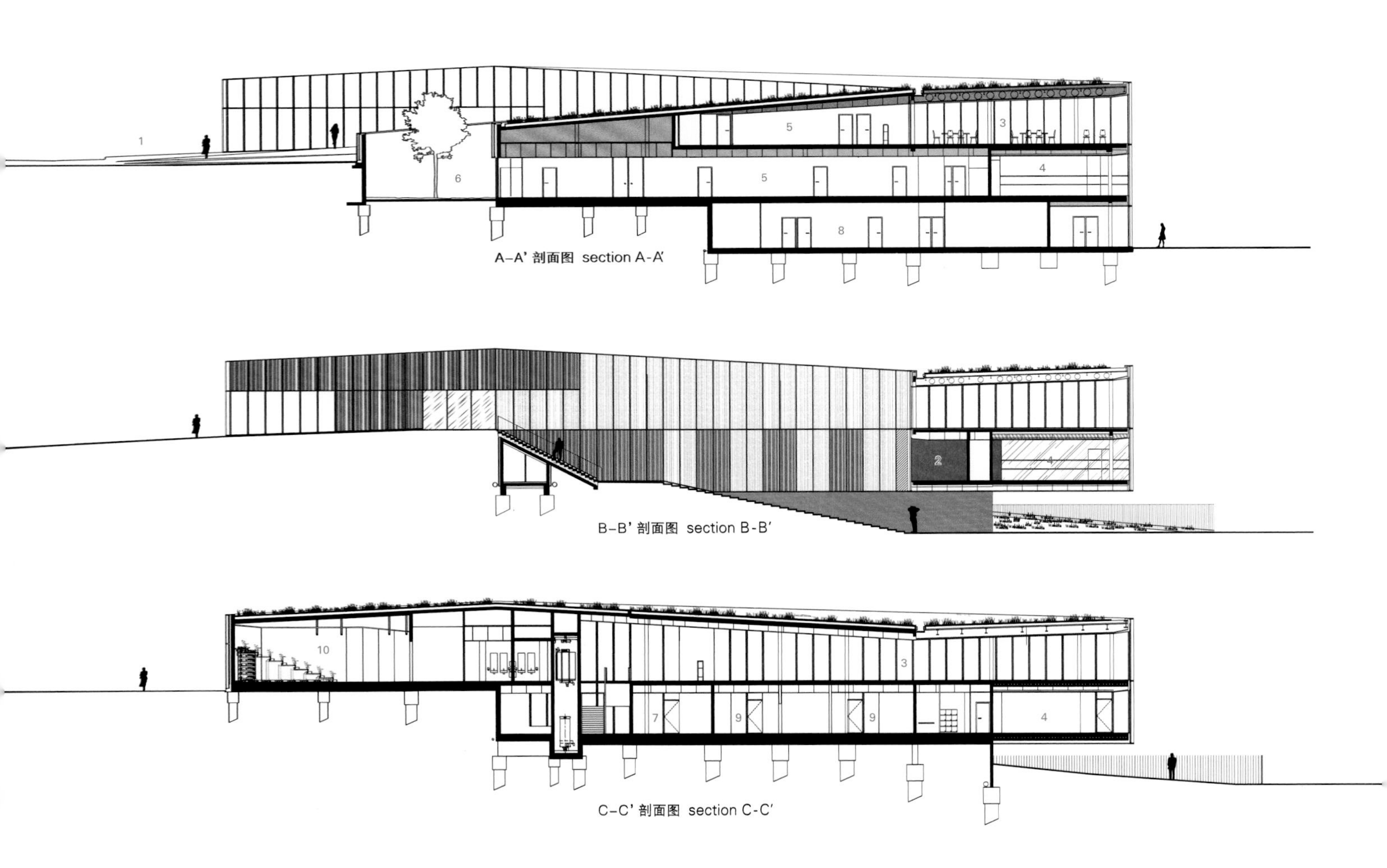

A-A' 剖面图 section A-A'

B-B' 剖面图 section B-B'

C-C' 剖面图 section C-C'

Technical premises and bicycles parking are provided at the lower public square.
At the lower square, the building is dug in to leave the upper part of the building in levitation. The library emerges and raise thank to an impressive pre-stressed concrete beam which is set against a massive buried volume to balance the composition.
The west and east facades are offered one to the city and the other one at the valley L'Orge. The west facade is a signal for the higher square. The East facade frames the landscape and dominates the lower square searching of the horizon.
By a measured set of transparency, the new public building is a showcase about the city and the landscape allowing people firstly to see from outside its cultural activities and mostly to frame the landscape of the valley.
The design of the wall panels, based on a meticulous framework gradually alternating with different shaded of glazed panels and printed-panels, gives rhythm to the facade of the public building.
Solid wall-panels are made of extruded aluminum measured elements to create a random feeling and hide all mechanical fixations.
The roof is vegetated to increase significantly the inertia of the building and to preserve a pleasant views over the surrounding buildings.
This new building is a generator for a plot with great potential offering, the city of Saint-Germain-les-Arpajon the cultural and social urban center it deserves. The urban composition, a stretch line in the landscape, complemented by an clear architectural style, allows the public building to become an attraction in the neighborhood but also a link between the Leuville road and the valley L'Orge.
The project was designed according to a comprehensive though about location, shape, cost of operation and materials, aiming to promote exchange, conviviality and allow public meeting.

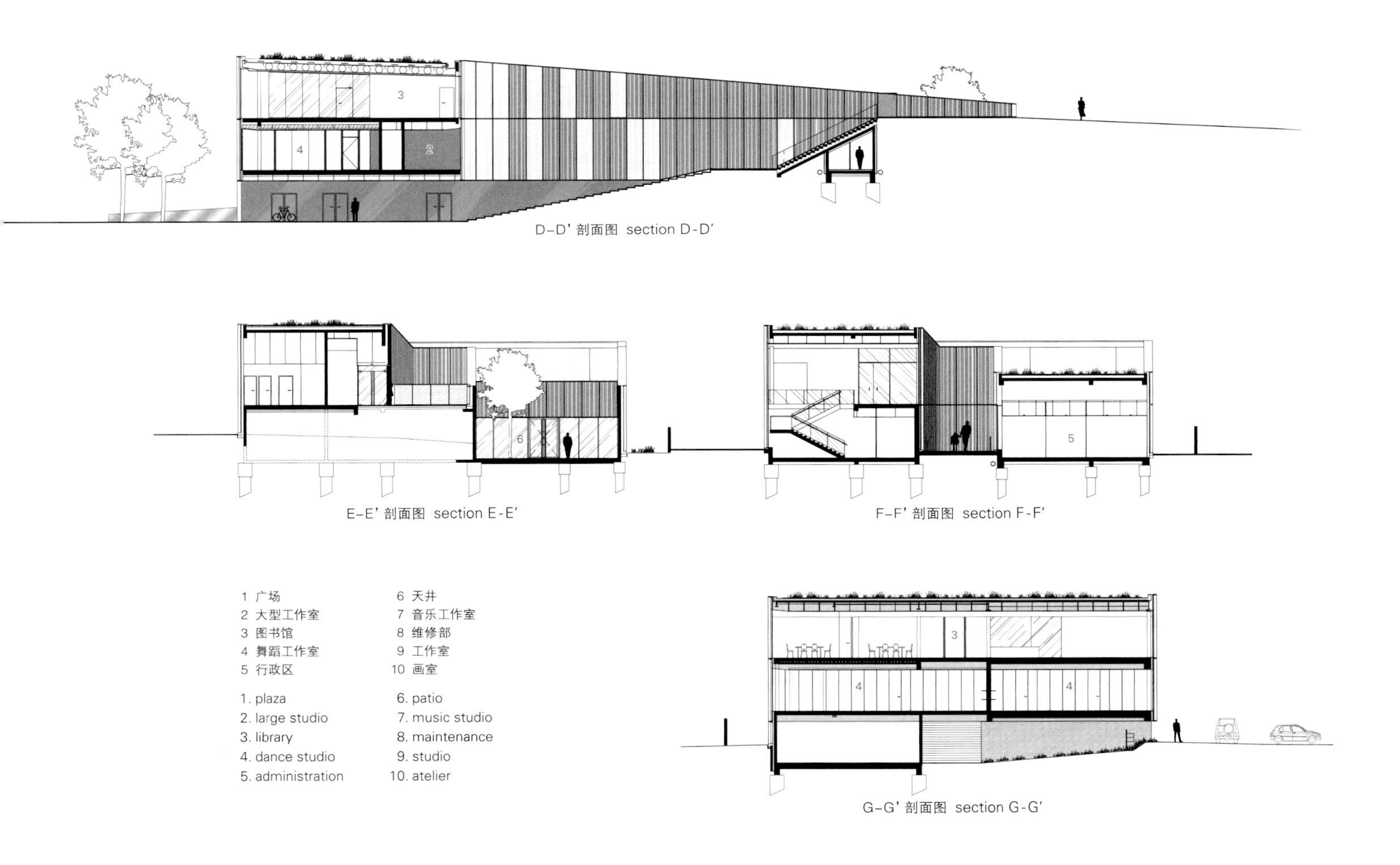

D-D' 剖面图 section D-D'

E-E' 剖面图 section E-E'

F-F' 剖面图 section F-F'

1 广场
2 大型工作室
3 图书馆
4 舞蹈工作室
5 行政区
6 天井
7 音乐工作室
8 维修部
9 工作室
10 画室

1. plaza
2. large studio
3. library
4. dance studio
5. administration
6. patio
7. music studio
8. maintenance
9. studio
10. atelier

G-G' 剖面图 section G-G'

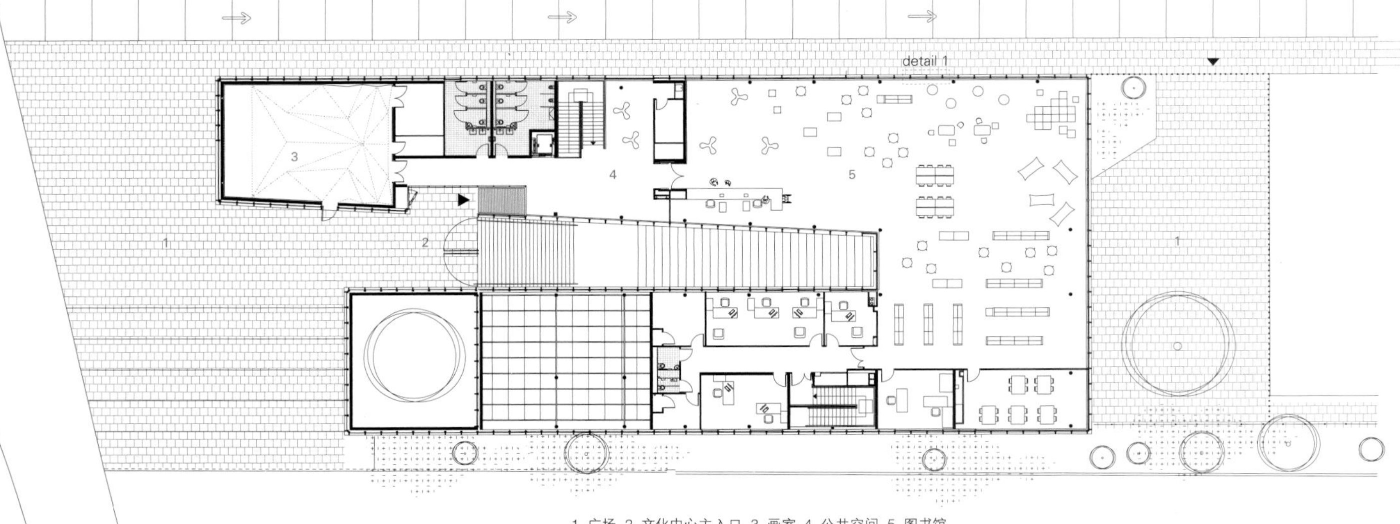

1 广场 2 文化中心主入口 3 画室 4 公共空间 5 图书馆
1. plaza 2. cultural center's main entrance 3. atelier 4. common space 5. library
二层 first floor

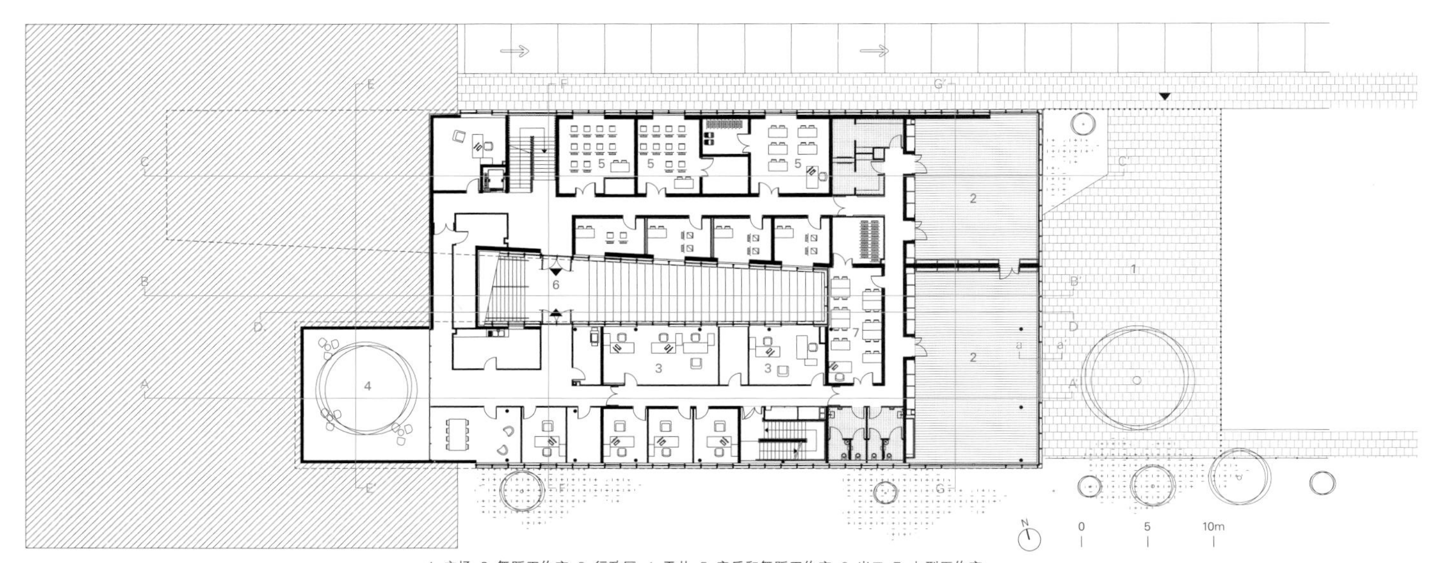

1 广场 2 舞蹈工作室 3 行政区 4 天井 5 音乐和舞蹈工作室 6 出口 7 大型工作室
1. plaza 2. dance studio 3. administration 4. patio 5. music and dance studio 6. exit 7. large studio
一层 ground floor

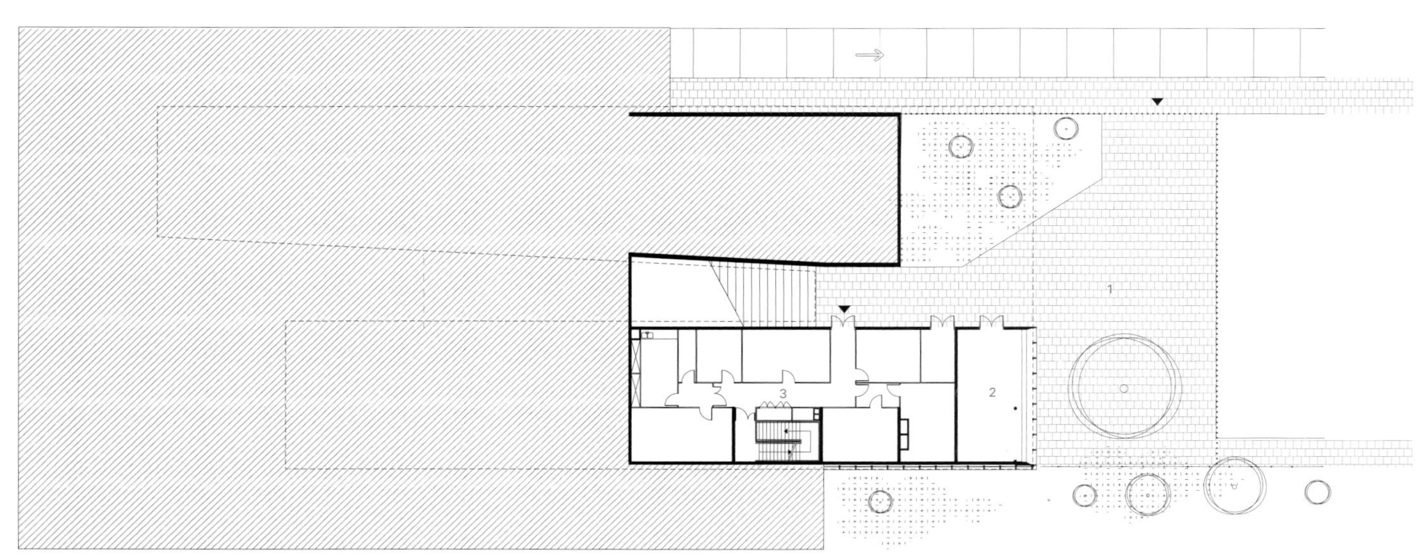

1 广场 2 自行车存放处 3 维修部
1. plaza 2. bicycle 3. maintenance
地下一层 first floor below ground

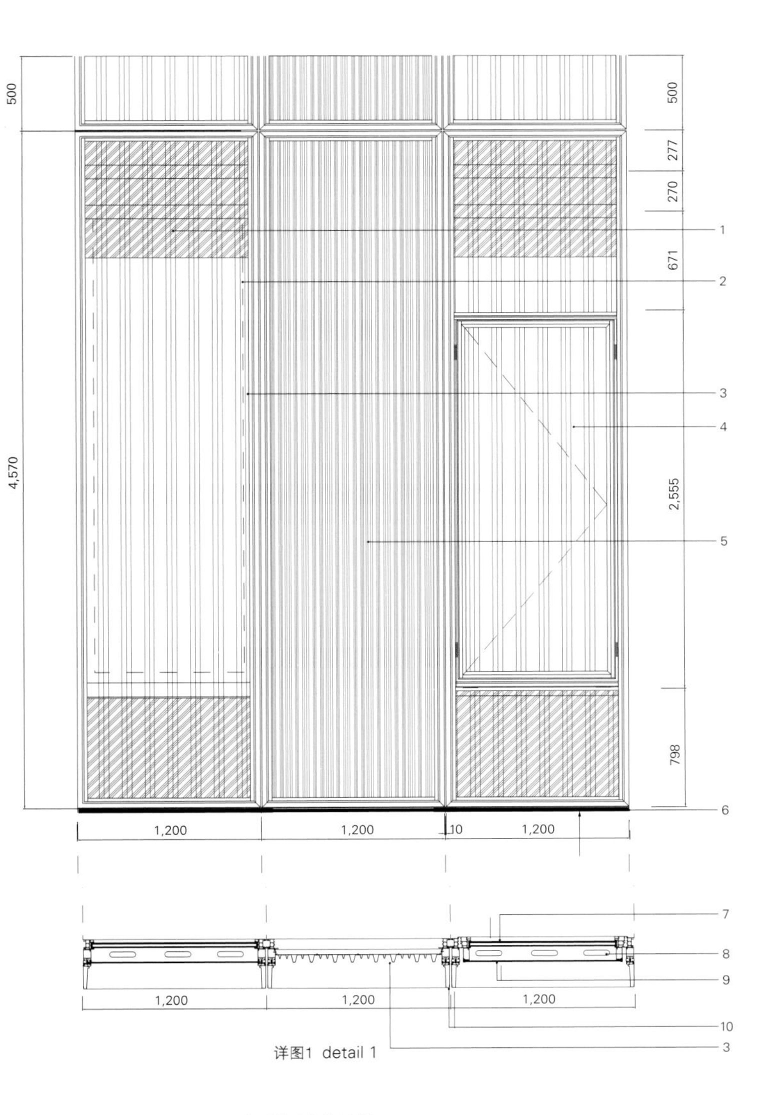

详图1 detail 1

1. hidden sheet metal for blind RAL9006
2. mechanical blind RAL9006
3. extruded aluminium mesh interpon jaisalmer
4. breathing glass windows
5. isolated box made of extruded aluminium profile + mineral wool
6. sheet metal ceiling
7. isolated double glazing cristal clear low emission
8. ventilation grid mosquito repellent
9. simple glaze screen printed RAL7047
10. extruded aluminium profile interpon jaisalmer

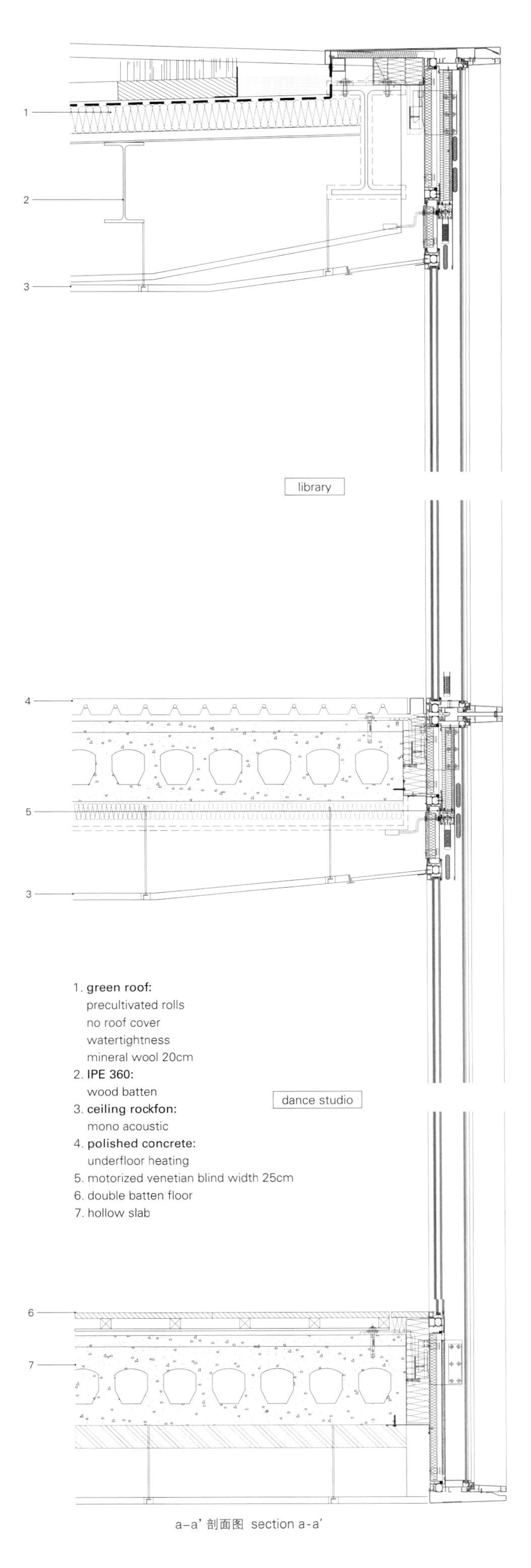

1. **green roof:**
   precultivated rolls
   no roof cover
   watertightness
   mineral wool 20cm
2. **IPE 360:**
   wood batten
3. **ceiling rockfon:**
   mono acoustic
4. **polished concrete:**
   underfloor heating
5. motorized venetian blind width 25cm
6. double batten floor
7. hollow slab

a-a' 剖面图 section a-a'

>>124

Parc Architekten

Was founded by Barbara Poberschnigg[right] and Michael Fuchs[left] in 2005. Barbara Poberschnigg graduated from the School of Applied Art and painting in Innsbruck. She also received a master's degree in Entrepreneurship(MSc) from the University of Liechtenstein. Michael Fuchs studied political science and history at the University of Innsbruck. And he also studied architecture at the University of Innsbruck and National School of Architecture Paris La Villette.

>>10

Wilkinson Eyre Architects

Is one of the UK's leading architectural practices, based in London, England. Chris Wilkinson set up Chris Wilkinson Architects in 1983 and formed a partnership with Jim Eyre in 1987, with the practice registered as Wilkinson Eyre Architects in 1999. His contribution has been recognized by the award of an OBE(Officer of the Order of the British Empire) in the Millennium Honors List, election to the Royal Academy of Arts in 2006, and an Honorary Fellowship of the AIA in 2007. He combines a life-long interest in art with a fascination for science, technological innovation and a sense of history, producing a fresh new approach to architecture. Is the only practice in the UK to have had back to back success for both the RIBA Stirling Prize and the RIBA Lubetkin Prize.

>>64

Elemental

Is a group of architects, founded in 2001 and led by five partners Alejandro Aravena, Gonzalo Arteaga, Juan Ignacio Cerda, Diego Torres and Victor Oddó[from the left].

Alejandro Aravena, CEO of Elemental, is a member of the Pritzker Prize Jury since 2009. He was named Honorary RIBA International Fellow in 2009. He is the Director of 15th Venice Biennale International Architecture Exhibition 2016. Elemental is a "Do Tank" company, the strength of which is the innovation and design quality of public interest and social impact, with a team highly enabled in the development of complex initiatives, which require the coordination of public and private actors and participating processes of decision; working in urban projects of infrastructure, public space, transportation and housing, operating within the city and its capacity to generate wealth and quality of life, using it as a shortcut towards equality.

>>22

JCY Architects and Urban Designers

Is an internationally recognized Western Australian owned architectural practice focused on providing the highest quality built-form outcomes for a broad range of clients within both the public and private sectors. They have produced sustainable and innovative solutions for an extensive range of projects of varying architectural typologies that eschew a trademark style or response. They seek to provide a unique and appropriate design response to every project, with livability, longevity, durability and excellent design being of paramount importance. They have worked in urban, rural and remote areas of Australia as well as in Malaysia and New Zealand and Japan, where its skills and design excellence have been sought by numerous clients.

©Kim Courrges

©Brigitte Lacombe

>>50

Zaha Hadid Architects

Zaha Hadid is an architect who consistently pushes the boundaries of architecture and urban design. She studied architecture at the Architectural Association in 1972. Taught Architectural Design at Yale University as a visiting professor. Is currently a professor at the University of Applied Art in Vienna, Austria. Her interest lies in the rigorous interface between architecture, urbanism, landscape and geology as her practice integrates natural topography and human-made systems, leading to innovation with new technologies.

>>36

John Wardle Architects

The design process of JWA emphasizes ideas that often traverse across the diverging scales and project types undertaken by the practice. Narrative is employed by JWA as a device for navigating through the development of a design using the personal and collective histories of the practice's clients, memories associated with a particular site, landscape and built context as generators of a cohesive narrative. In 2002 and 2006, JWA was awarded the Royal Australian Institute of Architects Sir Zelman Cowen Award for the most outstanding work of public architecture in Australia. The practice has also been awarded the Harold Desbrowe-Annear Residential Award on three occasions, the Victorian Architecture Medal for a second time in 2008 and the Esmond Dorney Award for Residential Architecture in 2012.

>>112

ZAO / Standardarchitecture

Was founded by Zhang Ke in 2001. Is a leading new generation design firm engaged in practices of planning, architecture, landscape and product design. Based on a wide range of realized buildings and landscapes in the past ten years, it has emerged as the most critical and realistic practice among the youngest generation of Chinese architects and designers. Zhang Ke received a B.Arch and M.Arch from School of Architecture at the Tsinghua University. In 1998, he received another M.Arch at the Graduate School of Design, Harvard University. Currently he lectures at various Universities including Tsinghua University and University of Hong Kong(HKU). Was awarded Design Vanguard of the World from the Architecture Record in 2010.

NADAAA

Is a Boston-based architecture and urban design firm led by principal designer Nader Tehrani, in collaboration with partners Katherine Faulkner and Daniel Gallagher. Is a platform for design investigation at a large scale and with a great geographic reach. The integration of landscape and building is part of their foundational standards, binding it to the design process, and finding innovative ways in which to intertwine it into systems and plans. Tehrani is also the Dean of the Irwin S. Chanin School of Architecture at the Cooper Union. He has taught at the MIT School of Architecture, where he served as the Head of the Department from 2010-2014. He has also taught at Harvard Graduate School of Design and Rhode Island School of Design.

Tehrani's work has been recognized with notable awards, including the Cooper Hewitt National Design Award in Architecture(2007), the United States Artists Fellowship in Architecture and Design(2007)

>>154

aplusarchitects

Was established in 2005 by Anna Mária Tamás and Krisztián Kovács-Andor, full time faculty at the University of Pécs(PTE). They believe that architecture and sustainability are inseparable. They have been specialized in architecture, heritage protection, urban design and solidarity in architecture. In 2013, they founded the Research Group for Solidarity in Architecture whose focus is to benefit communities worldwide through innovative architecture and design. They have more than 10 years architectural design and higher education teaching experiences.

>>86

Tank Architectes

Two partners Olivier Camus and Lydéric Veauvy met each other in the first year at Saint-Luc Tournai in 1992. They graduated with honors in 1999 and 1998 respectively and created Tank Workshop Architects in Lille in 2005. With their long experience of the house, they could explore all the typological variety of contemporary living.

项目名称：Ngoolark, ECU Student Services Building 34 / 地点：Joondalup, Perth, Western Australia
建筑师：JCY Architects and Urban Designers / 设计建筑师：Libby Guj / 项目建筑师：Will Thomson
项目团队：Paul Jones, Glenn Russell, Madeleine Hug, Jason Welten, James Bolger, Clare Porter
项目经理：NS Projects-Stewart Greensmith & Richard Yeoh / 结构、土木工程师：BG&E
机械和电气工程师：Wood & Grieve Engineers / 水力和消防工程师：SPP Group
景观设计师：Plan E Landscape Architects / 音效设计师：Gabriels Environmental Design / ESD设计师：Umai Low
安全高度设计：Altura / 立面顾问：Arup
艺术家：Andrew Stumpfel and Sohan Ariel-Hayes / 公共艺术协调师：Andra Kins
承包商：PACT Constructions-Kelvin Chance _ Project manager, Steve Ball _ Construction manager
现场监控师：Peter Turner, Simon Grant, Steve Hearne, Bevan Scanlon, John Lyons
甲方：Edith Cowan University
用地面积：9,342m² / 总建筑面积：2,300m² / 有效楼层面积：12,750m²
设计时间：2011 / 施工时间：2013—2015
摄影师：
©Peter Bennetts (courtesy of the architect) - p.22~23, p.25, p.30, p.31, p.32, p.41
©Rob Ramsay (courtesy of the architect) - p.26~27, p.33

courtesy of the architect

墨尔本大学的墨尔本设计学院

John Wardle Architects + NADAAA

墨尔本大学墨尔本设计学院的新大楼是为建筑建设和规划学部的师生设计的，以鼓励该学部的师生与其他学部合作，共同开展研究项目。新大楼共有六层，包括一系列工作室、一个工作室大堂和中庭、报告厅、一座图书馆、展览空间、专家车间，另外它还整合了两个具有重大历史意义的元素：一个是建于1856年的新南威尔士银行的外立面，另一个是日式房间。

新大楼作为本科教育和研究生教育基地的枢纽，通过其外部结构和内部设计，使学生能够对建成环境方面可持续发展的问题进行重要的跨学科研究探索。这栋建筑本身就是一个进行试验和研究的实验室。

墨尔本设计学院位于墨尔本大学帕克维尔校区历史核心的中心，其大楼整体全方位地与历史建筑、景观和主要的校区街道景观相得益彰。通过营造城市空间，每个立面及其相关的邻接处理设计都与周围具体的环境特点遥相呼应。而这些城市空间的设计则回应到了场地的东面和西面原有的阅览区，以及建筑物周围的开放性，来鼓励人们在此驻足、互动，突出行人的流动性。

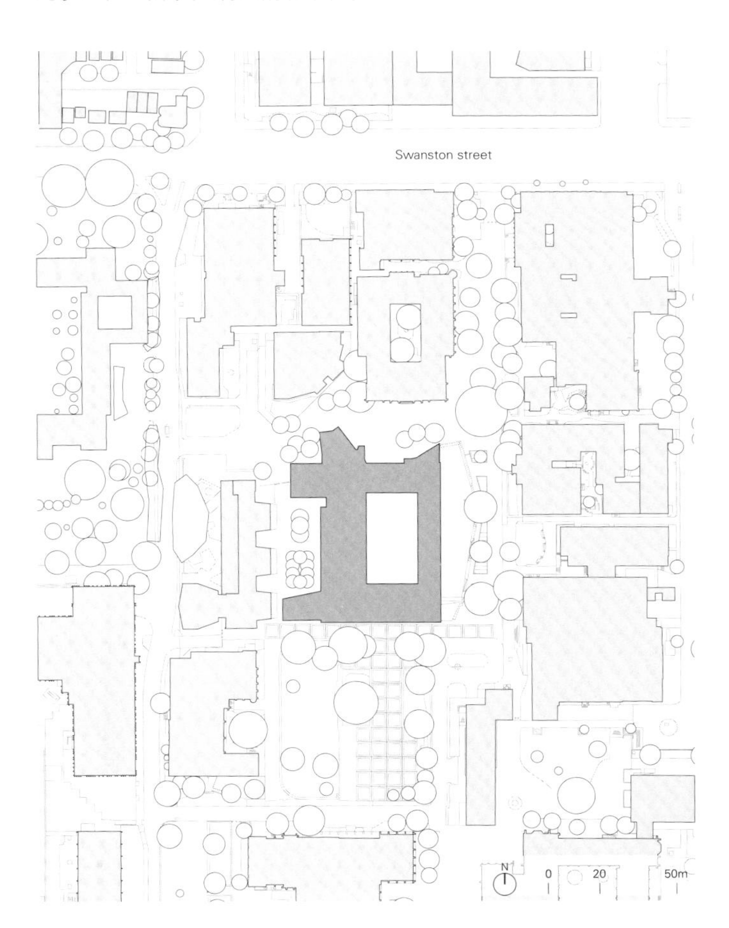

Melbourne School of Design
Melbourne School of Design

透明宽敞的一楼将该建筑向路人开放，进一步加强了大楼与更大范围的校园社区之间的联系。

该建筑的主要设计理念是：建筑教学法、工作室、学术环境和居住建筑。该建筑作为一个教学工具，通过其实际运作和建筑本身，很好地解释说明了各施工层的逻辑关系。建筑体系是暴露的，施工层是可拆分的，而像窗户和隔断这样的元素是可操作的。

在探索建筑教学法方面，通过其结构布局、材料组成、用户体验、结构和服务系统，该建筑向人们揭示了建筑形式和建筑物具有传授和"制造知识"的能力。

行政办公室、教职员工办公室和教学工作室的空间布局相互交织，使建筑学科、建筑物、城市主义和景观建筑之间的接触更加密切，且相互融合。

新的学术环境提供了一系列工作空间，鼓励伙伴们开展协作与对话。整栋建筑就是一个大型社区，而不同楼层又把学术工作空间划分出一个个小"村庄"。

通过使用一系列被动式设计原则、被动式太阳能设计、外立面太阳能响应系统、混合模式的自然通风、自然对流通风、车间大厅内被动热通风堆栈、可再生的建筑材料，例如，能减少内含能的单板层积材横梁和花格镶板、锌屏，整体设计已经达到了六星级的绿星教育建筑级别。在委托人、承包人和顾问团队的密切合作下，该建筑在预算内提前四个月完工。

Melbourne School of Design at University of Melbourne

The new building for the Melbourne School of Design at the University of Melbourne has been designed to accommodate students and staff from the Faculty of Architecture Building and Planning and to encourage the development of research projects in collaboration with other Faculties. Occupying six levels the new facility includes a series of studio spaces, a studio hall and atrium, lecture theaters, a library, exhibition spaces, specialist workshop spaces and the integration of two significant historical elements, the 1856 Bank of NSW facade and the Japanese Room.

The building serves as a hub for undergraduate and graduate education, enabling critical interdisciplinary exploration into issues of sustainability in built environments through its external structure and internal design. The building itself is a laboratory for experimentation and research.

Located in the center of the historic core of the University's Parkville campus, the Melbourne School of Design has been created as a building in the round in response to surrounding heritage buildings, landscapes, and key campus streetscape. Each elevation and its associated programmatic adjacency respond to the specifics of their context through the creation of urban spaces that emphasize both occupation, interaction, and pedestrian movements, responding to historic bookends to the east and west of the site and the opening up of activities at the building perimeter.

A transparent ground plane opens up the building to pass-

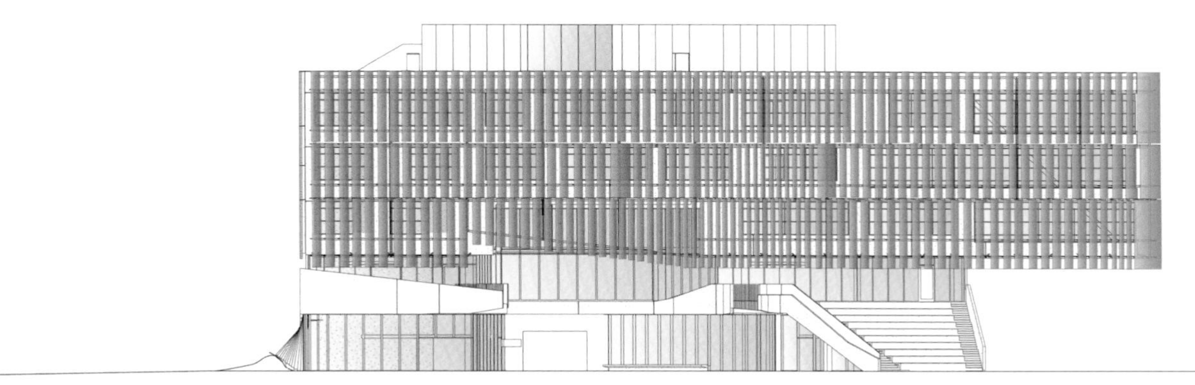

东立面 east elevation

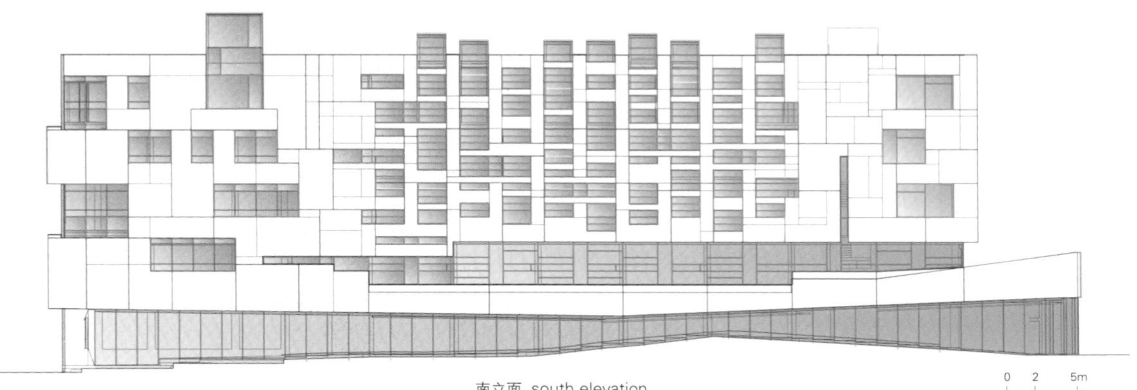

南立面 south elevation

项目名称：Melbourne School of Design at University of Melbourne / 地点：University of Melbourne, Parkville campus, Melbourne, Australia
建筑师：John Wardle Architects + NADAAA / 主要负责人：John Wardle, Stefan Mee _ John Wardle Architects, Nader Tehrani _ NADAAA
高级经理：Meaghan Dwyer / 项目建筑师：Stephen Georgalas / 项目经理：John Chow / 项目团队：Arthur Chang
室内设计师：Bill Krotiris, Andy Wong, Jasmin Williamson, Adam Kolsrud, Alex Peck, Barry Hayes, Jeff Arnold, Amanda Moore, James Loder, Danny Truong, Stuart Mann, Meron Tierney, Kenneth Wong, Sharon Crabb, Yohan Abhayaratne, Rebecca Wilkie, Ben Sheridan, Giorgio Marfella, Kirrilly Wilson, Elisabetta Zanella, Goran Sekuleski, James Stephenson, Adrian Bonaventura, Genevieve Griffiths, Michael Barraclough, Matthew Browne, Maria Bauer, Anja Grant _ John Wardle Architects, Katie Faulkner, James Juricevich, Parke MacDowell, Marta Guerra Pastrián, Tim Wong, Ryan Murphy, Ellee Lee, Kevin Lee, Rich Lee _ NADAAA
结构和土木工程师：John Wardle Architects, NADAAA / 机械和电气工程师：Irwin Consult / 机械、电气工程师：Aurecon
地热设计师：Douglas Partners Pty Ltd / 工料测量师：Rider Levett Bucknall / 建筑服务工程、可持续性顾问：Umow Lai
建筑可持续性委托顾问：AG Coombs / 建筑认证：Mckenzie Group
流线设计顾问：One Group ID Consulting / 音效顾问：Aecom / 景观建筑师：Oculus
照明设计师：Electrolight / 遗产地建筑师：RBA Architects, Conservation Consultants / A&V设计师：AVDEC / 交通工程设计师：Cardno Melbourne
面积：15,772m² / 造价：129,350,000 / 竣工时间：2014
摄影师：©Peter Bennetts-p.36~37, p.41, p.44, p.45, p.47, p.48, p.49 / ©Nils Koenning-p.39, p.42, p.43

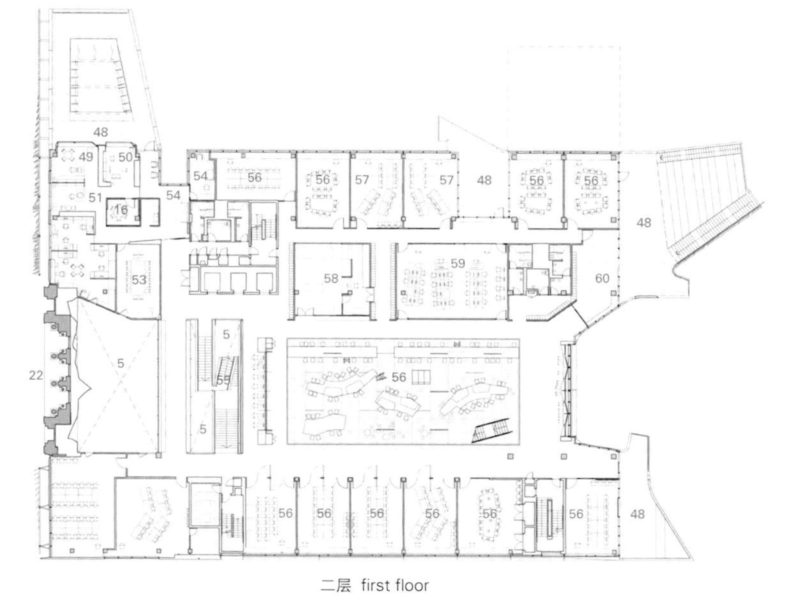

二层 first floor

五层 fourth floor

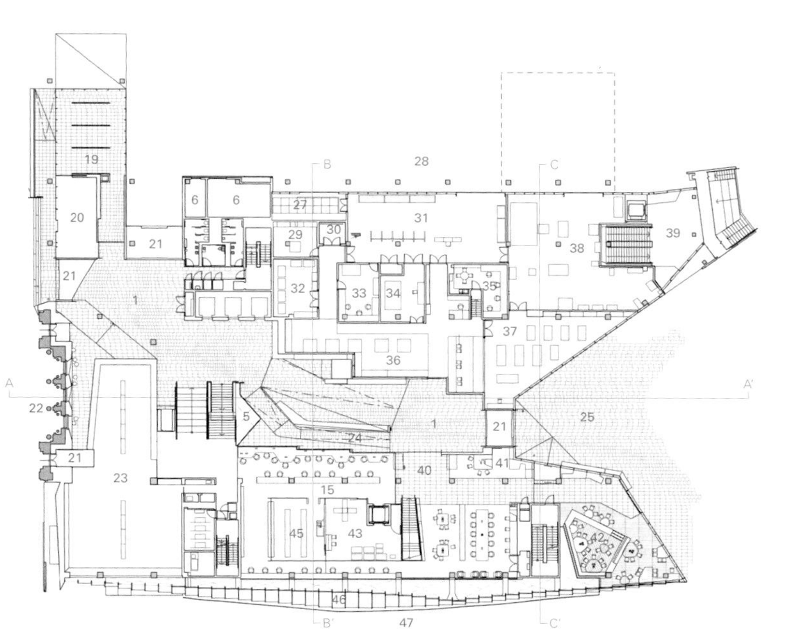

一层 ground floor

四层 third floor

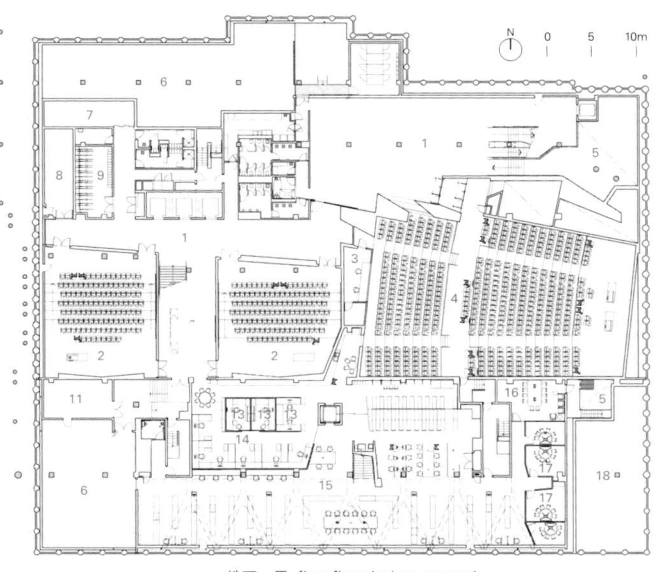

地下一层 first floor below ground

1 门厅
2 中型演讲厅
3 生物试验室
4 大型演讲厅
5 上空
6 发电室
7 储藏室
8 展览品储藏室
9 员工自行车储藏室
10 设计工作室
11 机键室
12 员工休息室
13 图书馆员工办公室
14 图书馆员工工作区
15 图书馆藏品区
16 会客室
17 图书馆放映室
18 水箱
19 小型展览区
20 咖啡室
21 入口气闸
22 遗迹立面
23 大型展览区
24 河床
25 入口庭院
26 日式露台
27 车间湿区
28 车间室外空间
29 金属车间
30 喷漆室
31 车间储藏室
32 激光切割室
33 数控车间
34 数控雕刻室
35 车间办公室
36 模型制作室
37 车间长椅区
38 机器车间
39 大型演讲厅的门厅
40 图书馆入口
41 图书馆服务台
42 图书馆的协同工作区
43 还书室
44 日式房间
45 高频率展示区
46 图书馆上方的天窗
47 景观坡台
48 露台
49 院长办公室
50 副院长办公室
51 主任办公室
52 家具储藏室
53 会议室
54 非正式会面区
55 Y形楼梯
56 设计工作室
57 CAD工作室
58 打印室
59 CAD实验室
60 气闸
61 开放的学术车间
62 封闭的工作区
63 工作区
64 公共设施区
65 小厨房
66 教授工作区
67 RHDS休息室
68 RHDS学习区

1. foyer
2. medium lecture theater
3. bio box
4. large lecture theater
5. void to below
6. plant room
7. bin store
8. exhibition store
9. staff bike store
10. design studio
11. switchroom
12. staff lounge
13. library staff offices
14. library staff workspaces
15. library collection
16. meeting room
17. library projecting rooms
18. water tank
19. small exhibition
20. cafe
21. entry airlock
22. heritage facade
23. large exhibition
24. creek bed
25. entry courtyard
26. Japanese terrace
27. workshop wet space
28. workshop external space
29. metal workshop
30. spray booth
31. workshop storage
32. laser cutting room
33. digital control workshop
34. CNC router
35. workshop office
36. model making space
37. workshop bench space
38. machine workshop
39. large lecture theater's foyer
40. library entry
41. library service desk
42. library collaborative space
43. returns room
44. Japanese room
45. high use collection
46. skylight to library below
47. landscaped berm
48. terrace
49. dean office
50. deputy dean office
51. general manager office
52. furniture storage
53. conference room
54. informal space
55. Y-stair
56. design studio
57. CAD studio
58. print room
59. CAD lab
60. airlock
61. open academic workspace
62. enclosed workspace
63. workspace
64. utility
65. kitchenette
66. professor workspace
67. RHDS lounge
68. RHDS study space

ersby and encourages further engagement between the building and the broader campus community.

The major themes of the building brief were Built Pedagogy, the Studio, the Academic Environment and the Living Building. The building is explanatory in its operation and architecture, revealing a logic of construction layers as a pedagogical tool. Building systems are exposed, construction layers are pulled apart and elements like windows and partitions are operable.

In exploring Built Pedagogy the building reveals through its composition, material make-up, experience, structural and services systems, that built form and matter have the ability to teach and to "produce knowledge".

The spatial layout of the administrative and faculty offices and the teaching studios is interwoven, bringing the disciplines of architecture, building, urbanism, and landscape architecture into a more active contact, and collaborative participation with each other.

The new academic environment encourages collaboration and dialogue between peers, providing a range of workspace. The floor plate profile creates small villages of academic workspace within the overall community of the building.

The design has achieved a 6-Star Green Star Education rating (design) through the incorporation of good passive design principles, passive solar design, solar responsive facade systems, mixed mode natural ventilation, cross flow ventilation, a passive thermal ventilation stack within the studio hall, and the use of renewable building materials such as LVL beams and coffers and zinc screens resulting in reduced embodied energy. In close collaboration with the Client, the Contractor and the Consultant team the building was delivered four months ahead of schedule and on budget.

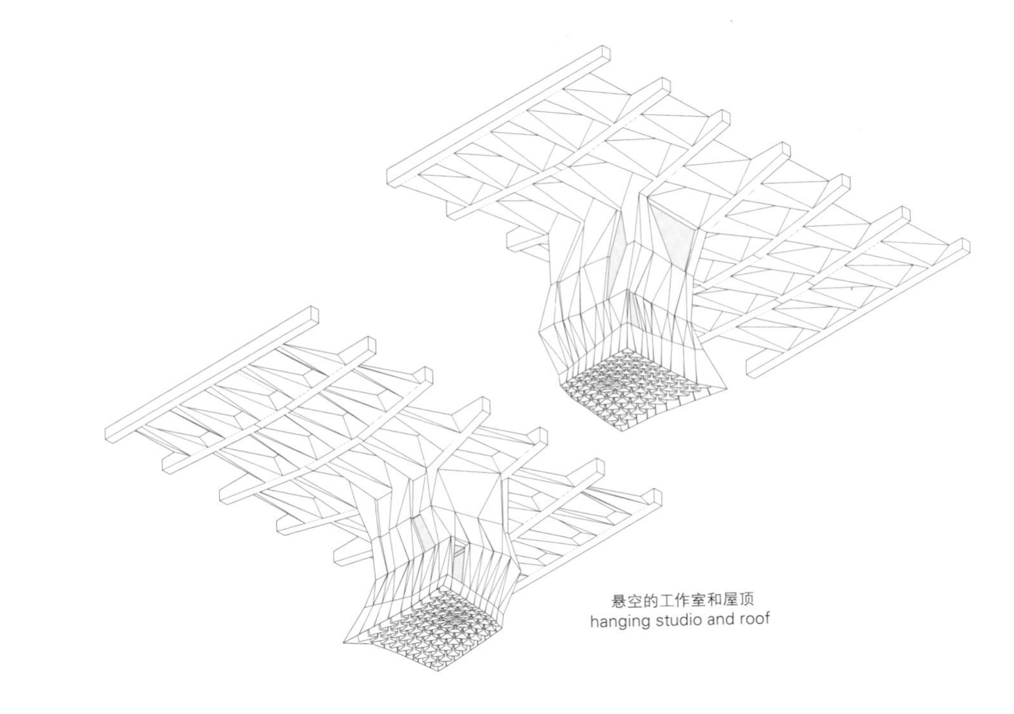

悬空的工作室和屋顶
hanging studio and roof

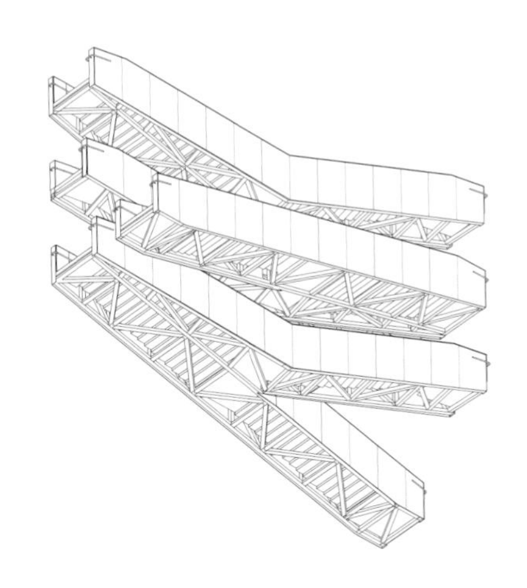

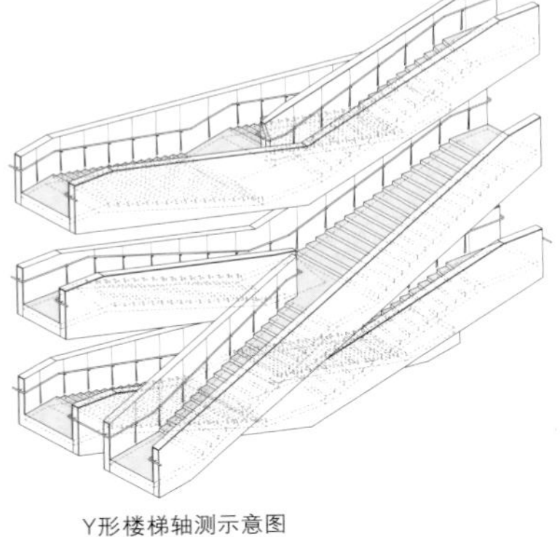

Y形楼梯轴测示意图
axonometric diagram of Y-stair

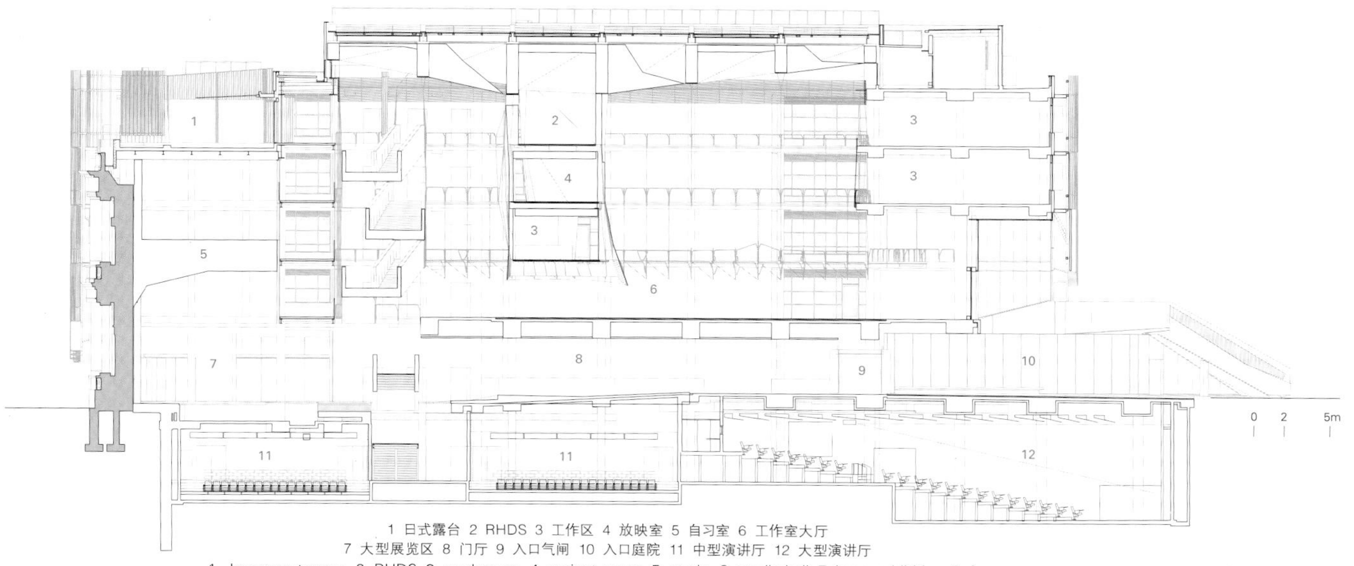

1 日式露台 2 RHDS 3 工作区 4 放映室 5 自习室 6 工作室大厅
7 大型展览区 8 门厅 9 入口气闸 10 入口庭院 11 中型演讲厅 12 大型演讲厅
1. Japanese terrace 2. RHDS 3. workspace 4. project space 5. study 6. studio hall 7. large exhibition 8. foyer
9. entry airlock 10. entry courtyard 11. medium lecture theater 12. large lecture theater
A–A' 剖面图 section A-A'

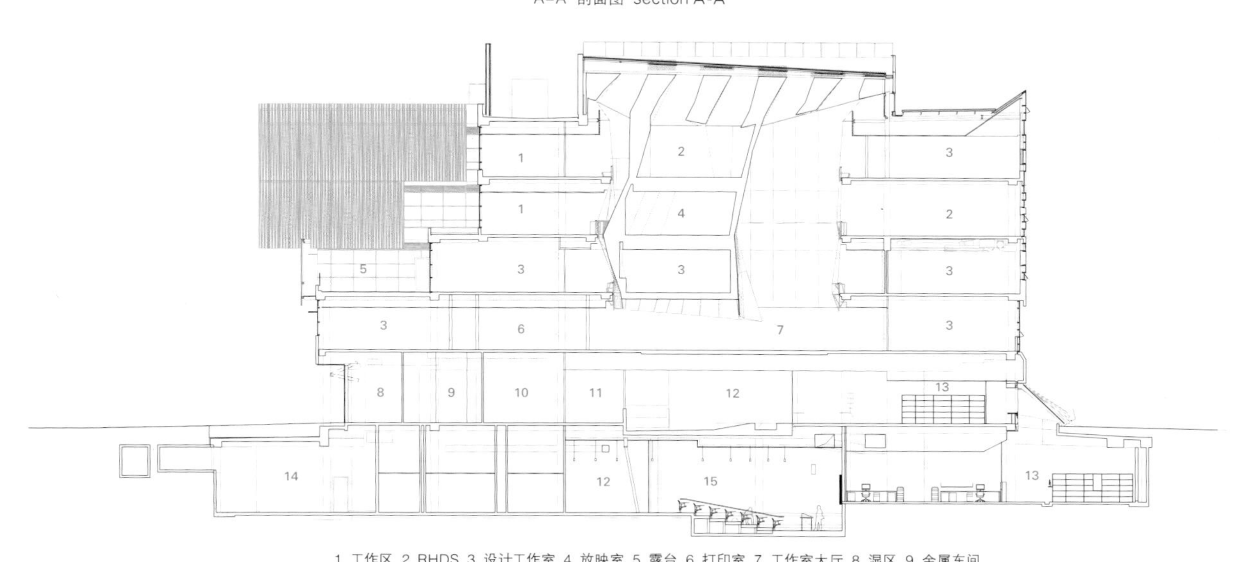

1 工作区 2 RHDS 3 设计工作室 4 放映室 5 露台 6 打印室 7 工作室大厅 8 湿区 9 金属车间
10 激光切割室 11 模型制作室 12 门厅 13 图书馆 14 发电室 15 中型演讲厅
1. workspace 2. RHDS 3. design studio 4. projecting space 5. terrace 6. print room 7. studio hall 8. wet space 9. metal workshop
10. laser cutting room 11. model making space 12. foyer 13. library 14. plant room 15. medium lecture theater
B–B' 剖面图 section B-B'

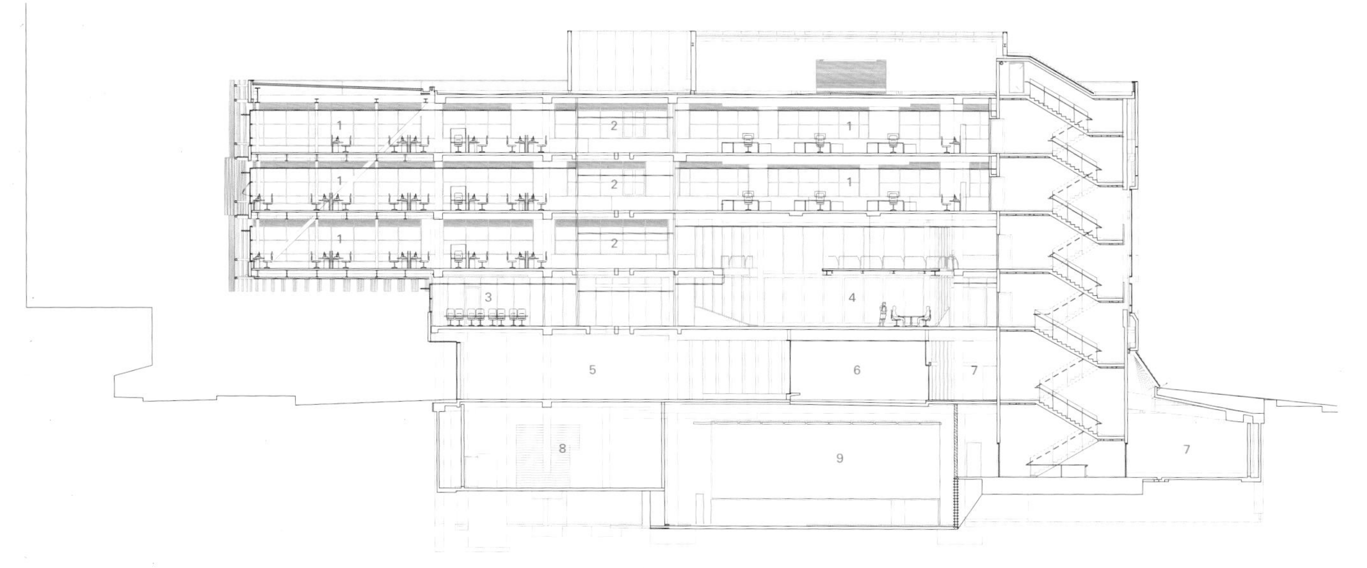

1 工作区 2 设备间 3 工作室 4 大厅 5 车间 6 入口 7 图书馆 8 门厅 9 大型演讲厅
1. workspace 2. amenities 3. studio 4. hall 5. workshop 6. entry 7. library 8. foyer 9. large lecture theater
C–C' 剖面图 section C-C'

THE ASS,
THE LION
AND THE FOX
THESIS STUDIO

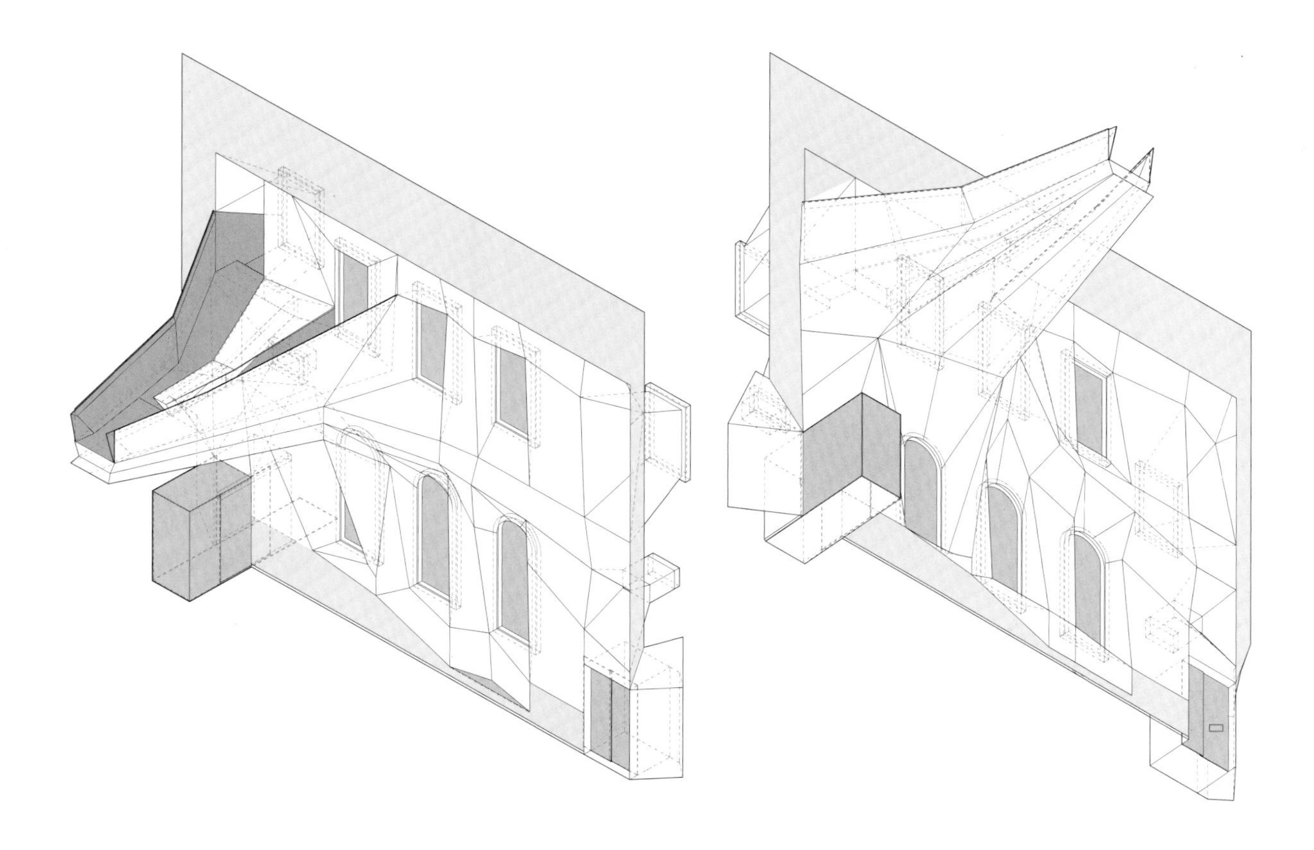

HOPPERS CROSSING STUDENT VILLAGE
SOCIAL INFRASTRUCTURE

牛津大学中东研究教学中心 Investcorp大楼

Zaha Hadid Architects

自1957年成立以来，牛津大学圣安东尼学院的中东研究教学中心一直致力于19世纪至今的阿拉伯世界、伊朗、以色列以及土耳其国家的研究和教学，其重中之重是人文与社会科学研究。从1978年至今，该中心就位于伍德斯托克路68号（此处曾是圣菲利普和圣杰姆斯教堂教区长的住宅，建于1887年）。中东研究教学中心拥有牛津大学关于现代中东地区研究最主要的收藏品，是世界级私人文件和历史照片档案馆，供对中东这一地区感兴趣的学者和科研人员查阅。该中心的研究焦点是专业图书馆、文献和照片档案，所收集的材料从19世纪开始至今。档案馆建于1961年，逐渐收集了400多个私人文件和十万多张历史照片。

中东研究教学中心Investcorp新楼为其增加了1127m²的占地面积和一间117人座位的阶梯教室，将该中心的图书馆及档案室的空间扩大了一倍，同时也为原存放于伍德斯托克路66号地下室里的藏品提供了更好的存储环境。

作为学院正在进行的扩建计划的一部分，Investcorp大楼融一些必要的新型设施于一体，以满足中东研究教学中心日益增长的研究活动和学术活动的需要。有了新的阶梯教室，中东研究教学中心就可以更多地开展深受欢迎的研讨会、讲座和辩论会。这些项目大部分都是向本校师生和公众开放的。

Investcorp大楼是对不断发展的学院的补充和完善。受圣安东尼学院里建筑地点的限制，Investcorp大楼蜿蜒迂回，与现有保护性建筑与树木相连并与之融为一体；其不锈钢外立面柔和地反射着自然光，与建筑周围景色遥相呼应。

该建筑项目受限于学院原有建筑和自然环境，但在设计中很好地整合了新增的学术和研究设施。这个项目保持了学院当前建筑所具有的独特个性，使其被解读为单一独特的元素，同时引入现代建筑，向人们传达了圣安东尼学院、牛津大学和这座城市从过去到现在和未来这一不断发展演变的趋势。

西侧，该建筑的规模受到了伍德斯托克路66号和68号现有建筑的限制。图书馆阅览室西侧外立面的弧形设计将千年的红杉树与其错综复杂的根部网络包含其中；基础板下面安装的是排水系统，确保红杉树可以吸收足够的水分。东侧档案阅览室和图书管理员办公室几乎与对面建于1970年的野兽派风格的希尔达–贝斯建筑的高度一致，但是

新Investcorp大楼仍然低于旁边伍德斯托克路66号建筑的屋顶线。

拥有117个席位的阶梯教室位于地下，通过错综复杂的管网通风换气。图书馆档案室下面还有一个类似的错综复杂的管网系统，作用是实现基本的环境控制，缓解机械调节空气流通的需求。地源热泵有效地控制着温度和湿度，为该中心著名的收藏品提供安全可靠的环境。

The Investcorp Building for Oxford University's Middle East Center

Founded in 1957, the Middle East Center at St Antony's College serves as the University of Oxford's facility for research and teaching on the Arab world, Iran, Israel and Turkey from the 19th century to the present day, with its focus on the research of humanities and social sciences. The center has been housed at 68 Woodstock Road (the former rectory of the

1 Investcorp大楼 2 伍德斯托克路66号 3 伍德斯托克路68号 4 牧师住宅 5 Hilda Besse大楼 6 Nissan中心 7 创始人大楼 8 门户大楼

1. Investcorp building 2. 66, Woodstock road 3. 68, Woodstock road 4. vicarage 5. Hilda Besse building 6. Nissan center 7. founder's building 8. gateway building

Church of St Philip and St James, built in 1887) since 1978. The Middle East Center holds Oxford University's primary collection on the modern Middle East, a world-class archive of private papers and historic photographs used by scholars and researchers with an interest in the region. The center's research core is the specialized library, document and photographic archive covering material from the 1800's onwards. The archive was set up in 1961 and has grown to over 400 collections of private papers and holds more than 100,000 historic photographs.

The new Investcorp Building for the Middle East Center provides 1,127 square meters of additional floor space and a new 117-seat lecture theater, doubling the space available for the Middle East Center's expanding library & archive, and providing optimum conditions to conserve and manage the center's collections that were previously stored in the basement of 66 Woodstock Road.

As an integral part of the college's on-going expansion plans, the Investcorp Building incorporates essential new facilities to meet the Middle East Center's increasing demand for research and academic activities. The new lecture theater will allow the Middle East Center to expand its popular program of seminars, lectures and debates – much of which is open to both the University and the general public.

The Investcorp Building complements the college's ongoing development. Its design weaves through the restricted site at St Antony's College to connect and incorporate the existing protected buildings and trees; while its stainless steel facade softly reflects natural light to echo the building's context.

The building integrates new academic and research facilities within a design defined by the existing built and natural environment of the college. The project maintains the detached character of the college's current buildings, allowing them to be read as separate elements, while introducing a contemporary building that conveys the past, present and future evolution of the college, university and city.

To the west, the project's scale defers to the existing buildings of 66 & 68 Woodstock Road. The curved form of library reading room's western facade accommodates the century-old Sequoia tree and its extensive root network; while a drainage system has been installed below the foundation slabs to ensure the tree receives enough moisture. To the east, the archive reading room and librarians' offices rise towards the height of the 1970 brutalist Hilda Besse Building it faces, yet the new Investcorp Building remains below the roofline of the adjacent 66 Woodstock Road.

The 117-seat lecture theater is located below ground and is ventilated through a labyrinth. A similar labyrinth exists beneath the library archive room to achieve the essential environmental controls and mitigate the need for mechanical air-handling. A ground source heat pump provides active ground coupling controlled for both temperature and humidity, creating a secure environment to conserve the center's renowned collection.

西立面 west elevation

南立面 south elevation

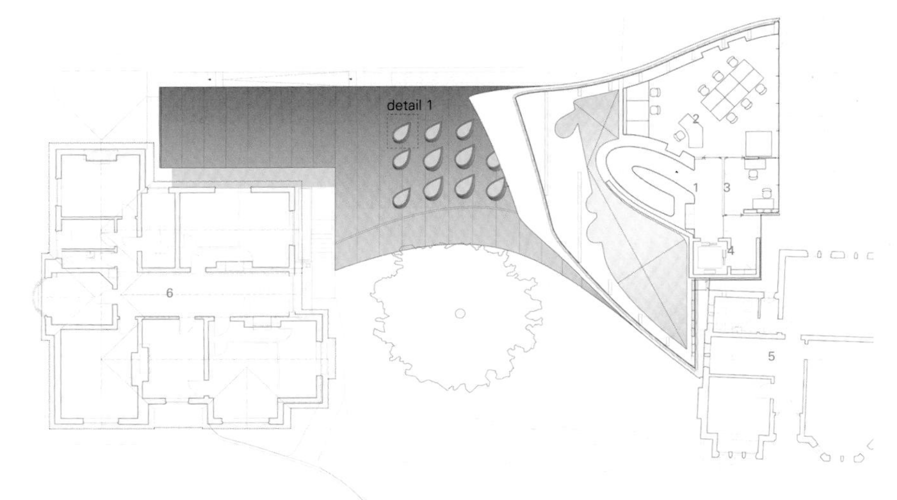

1 大厅 2 档案阅览室 3 档案办公室 4 电梯厅 5 伍德斯托克路68号 6 伍德斯托克路66号

1. hall 2. archive reading room 3. archive office 4. lift lobby 5. 68, Woodstock road 6. 66, Woodstock road

三层 second floor

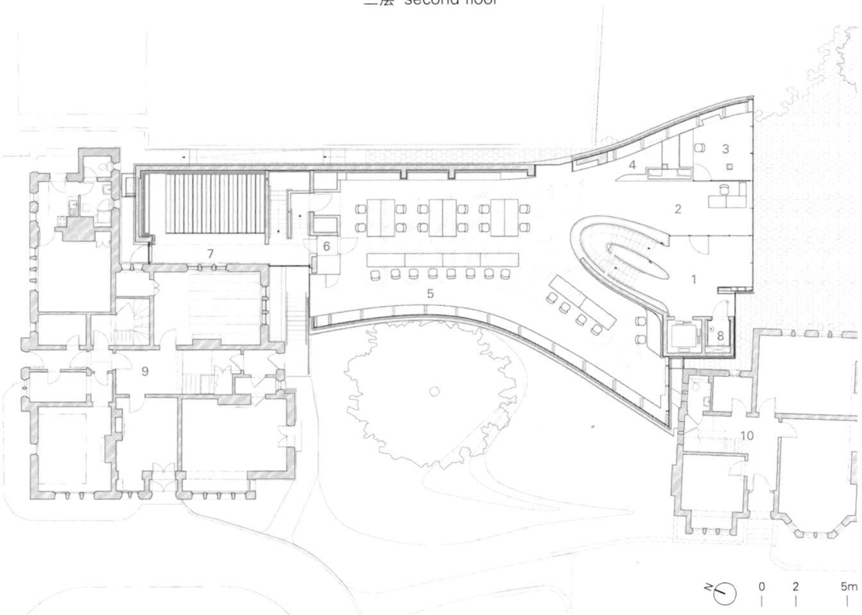

1 二层大厅 2 借还书处 3 图书馆管理员办公室 4 IT资源室 5 图书馆阅览区

6 图书升降区 7 图书馆可旋转的书架 8 卫生间 9 伍德斯托克路68号 10 伍德斯托克路66号

1. first floor hall 2. circulation desk 3. librarian's office 4. IT resource 5. library's reading room

6. book hoist 7. library's rolling stack 8. toilet 9. 68, Woodstock road 10. 66, Woodstock road

二层 first floor

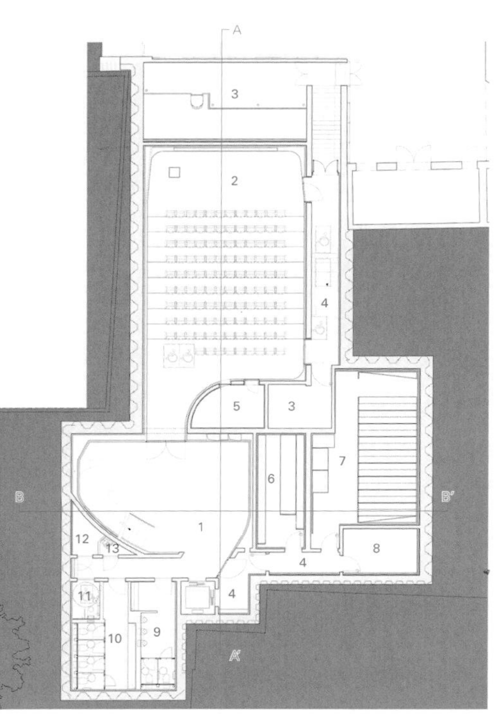

1 地下室大厅 2 讲厅 3 发电室

4 流线区 5 A&V室 6 照片档案室

7 文档档案室 8 新增资料区

9 男士卫生间 10 女士卫生间

11 变电室 12 服务间 13 清洁间

1. basement hall 2. lecture hall 3. plant room

4. circulation 5. audio & visual room

6. photo archive 7. paper archive

8. new accessions 9. male toilet 10. female toilet

11. electrical substation 12. server room

13. cleaner's storage

地下一层 first floor below ground

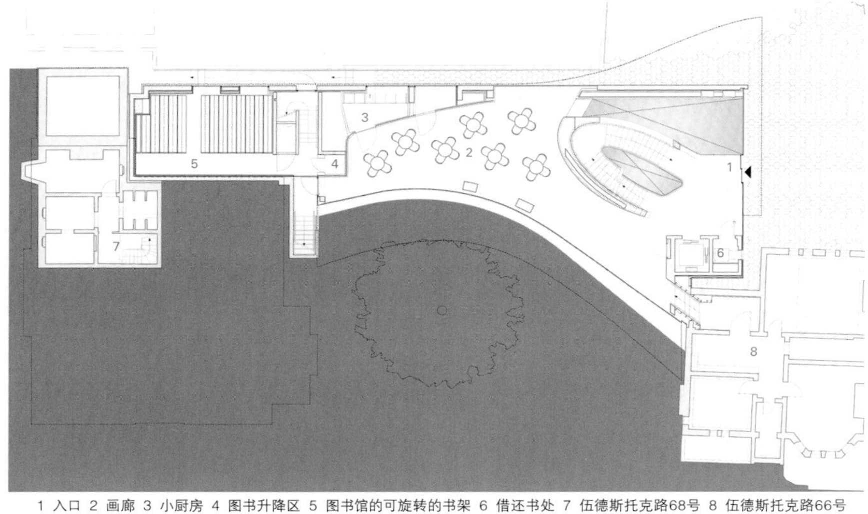

1 入口 2 画廊 3 小厨房 4 图书升降区 5 图书馆的可旋转的书架 6 借还书处 7 伍德斯托克路68号 8 伍德斯托克路66号

1. entrance 2. gallery 3. kitchenette 4. book hoist 5. library's rolling stack 6. book return

7. 68, Woodstock road 8. 66, Woodstock road

一层 ground floor

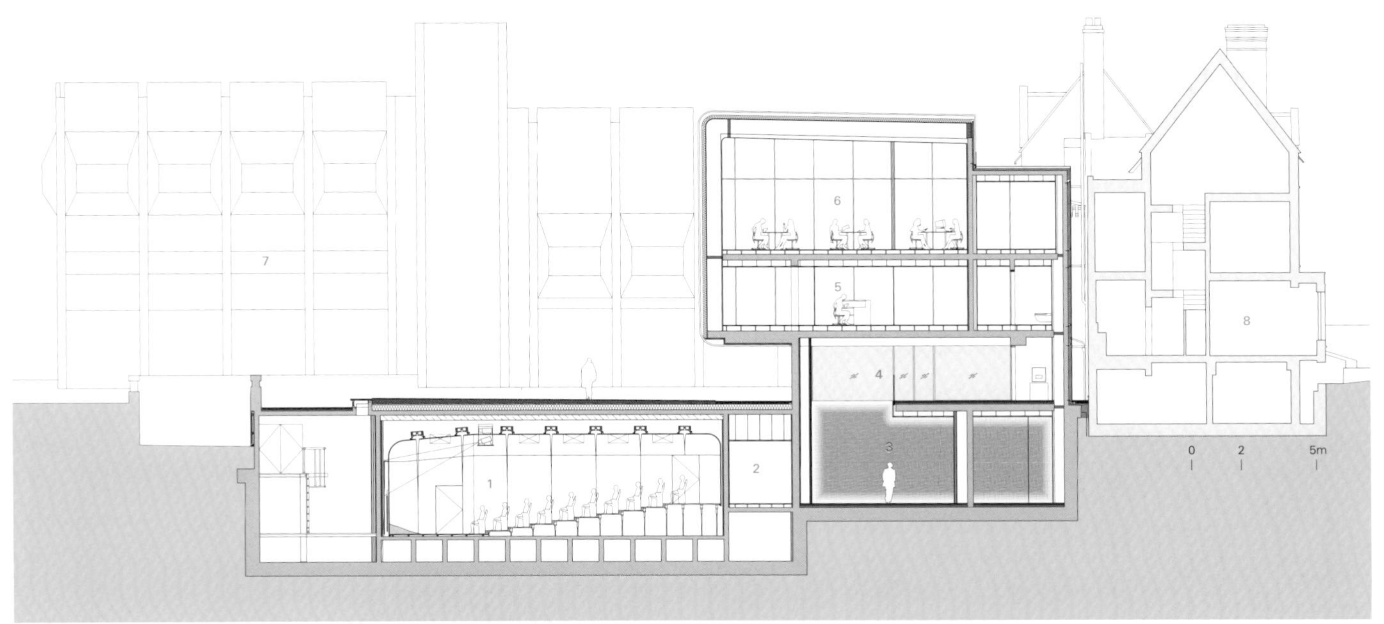

1 礼堂 2 A&V室 3 大堂 4 画廊 5 图书馆阅览区 6 档案阅览区 7 Hilda Besse大楼 8 伍德斯托克路66号
1. auditorium 2. audio & visual room 3. lobby 4. gallery 5. library's reading room 6. archive reading room 7. Hilda Besse building 8. 66, Woodstock road
A–A' 剖面图 section A-A'

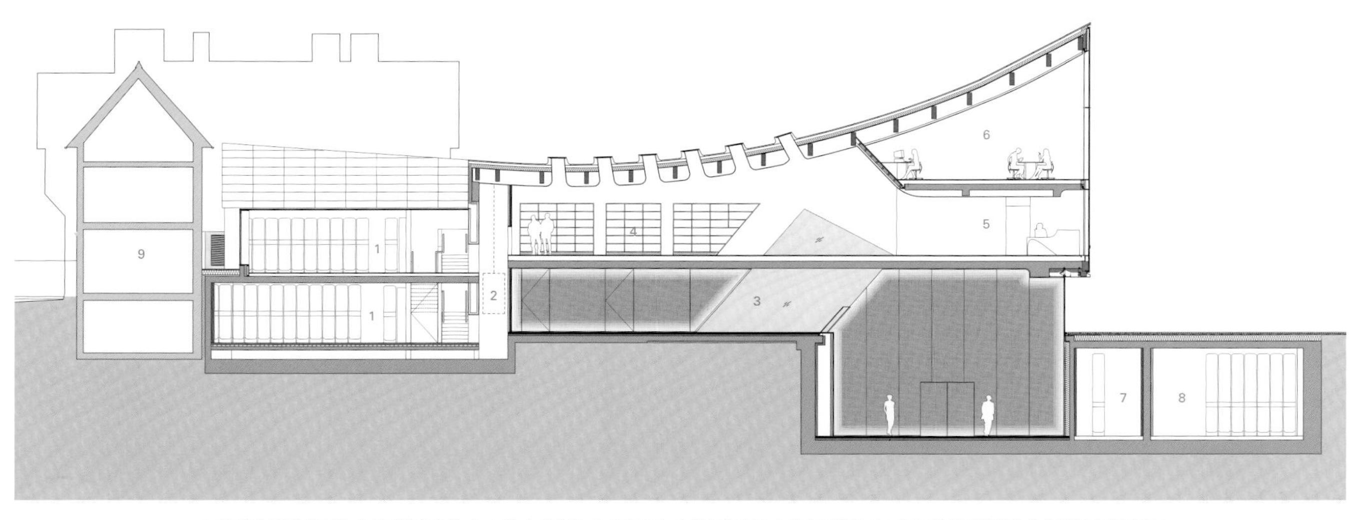

1 图书馆可旋转的书架 2 图书升降电梯 3 大堂 4 阅览室 5 借还书处 6 档案阅览室 7 照片储藏室 8 可旋转的档案储藏室 9 伍德斯托克路68号
1. library's rolling stack 2. book lift 3. lobby 4. reading room 5. circulation desk 6. archive reading room 7. photographic storage 8. rolling archive storage 9. 68, Woodstock road
B-B' 剖面图 section B-B'

项目名称：The Investcorp Building for Oxford University's Middle East Center at St Antony's College
地点：University of Oxford, Wellington Square, Oxford, OX1 2JD, United Kingdom
建筑师：Zaha Hadid Architects
设计师：Zaha Hadid
项目负责人：Jim Heverin
项目合作伙伴：Johannes Hoffmann; Ken Bostock
项目建筑师：Alex Bilton
项目团队：Sara Klomps, Goswin Rothenthal, Andy Summers, George King, Luke Bowler, Barbara Bochnak, Yeena Yoon, Saleem A Jalil, Theodora Ntatsopoulou, Mireira Sala Font, Amita Kulkarni
结构工程师：AKTII
机械/电气/音效工程师：Max Fordham
甲方的M&E顾问：Elementa
立面供应商：Frener + Reifer
立面顾问：Arup Facade Engineering
甲方的立面顾问：Eckersley O'Callaghan
承包商：BAM
项目经理：Bidwells
照明设计：Arup Lighting
造价顾问：Sense Cost Ltd.
消防工程师：Arup Fire
场地监督：Jppc Oxford
林业和树艺顾问：Sarah Venner
流线设计：David Bonnet
景观设计：Gross Max
CDM设计：Andrew Goddard Associates
视觉设计：Cityscape
甲方：Middle East Centre, St. Antony's College, University of Oxford, Eugene Rogan _ director
功能：archives, auditorium, rolling stacks, library's reading room, archive reading room, gallery
用地面积：1,580m²
有效楼层面积：1,127m²
设计时间：2006
施工时间：2013
竣工时间：2015
摄影师：©Luke Hayes (courtesy of the architect)

覆层详图
cladding detail

1. 10mm open joint
2. hairline joint
3. 2mm thick stainless steel rain screen panel with integral stiffeners
4. breathable membrane
5. insulation
6. polythene vapour control layer
7. 18mm plywood board fixed to glulam and tuner joist to support insulation
8. 150x63mm timber joists with packers/shaping to achieve final geometry
9. 165x450mm timber glulam beam
10. timber packer between glulam and plywood to achieve final geometry

详图1 detail 1

a-a' 剖面图
section a-a'

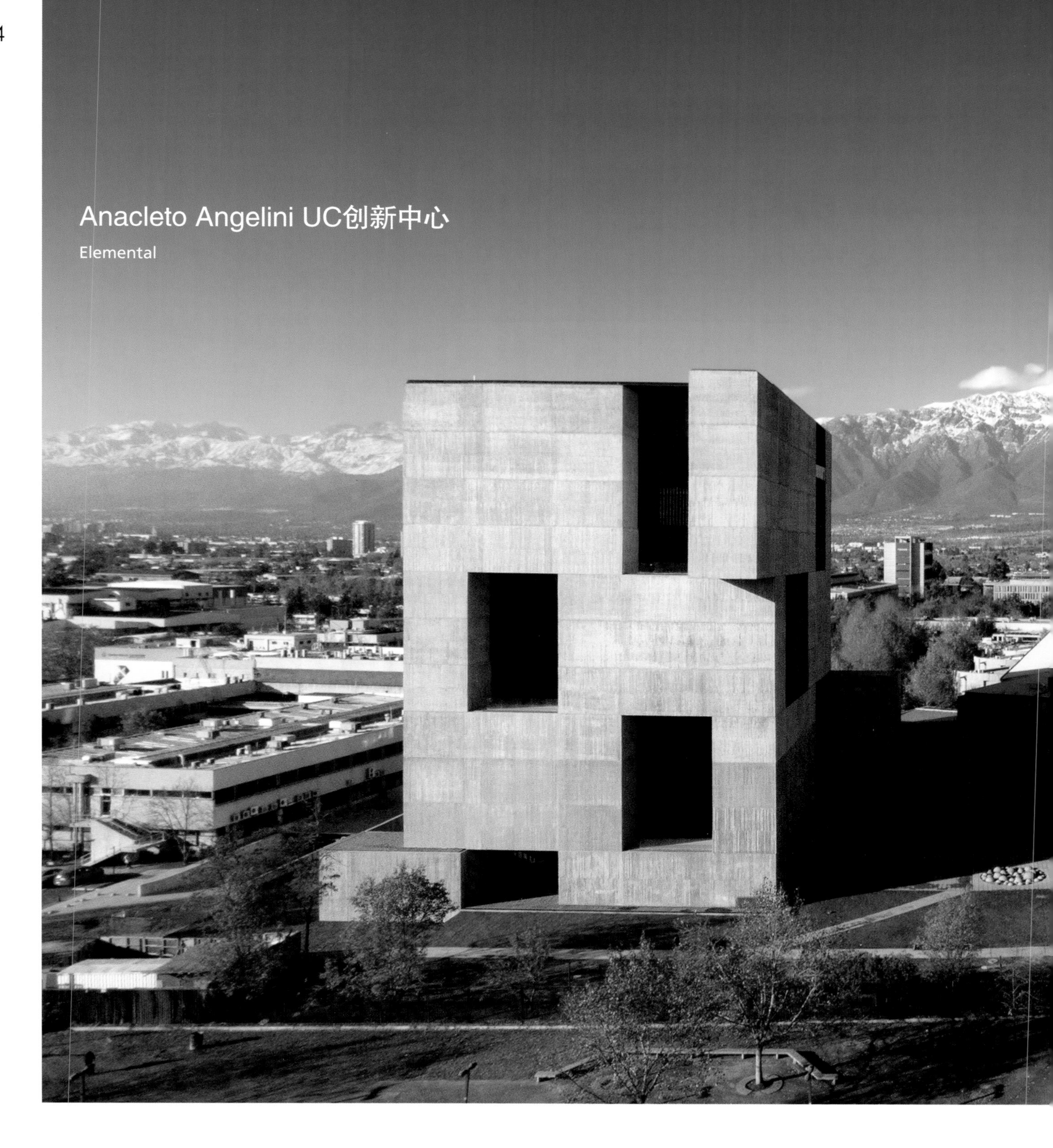

Anacleto Angelini UC创新中心

Elemental

2011年，Angelini集团决定捐资兴建一个创新中心，希望该中心能够将公司、企业和人们通常所说的需求与研究者和最先进的大学知识创造结合起来。其目的是加快技术转让过程，发现商业机会、提高现有资源利用的价值或注册专利，以提高国家竞争力，促进国家发展。这样的一个创新中心将在智利天主教大学的圣华金校区建造。

为了达成上述目标，我们认为创新中心的设计需要体现至少四种工作形式的交叉融合：正式工作形式和非正式工作形式，个人工作形式和集体工作形式。此外，我们认为，一个人要创造知识，面对面交流的方式是无与伦比的，所以该建筑里人们可以见面交流沟通的地方比比皆是：如果你打算乘电梯，碰巧遇到一个可以分享彼此感兴趣的信息的人，两人可以坐在电梯厅里的长椅上交谈；当你在大楼里直上直下移动时，透明的中庭会让你"偷窥"到别人正在做什么；且整个大楼自下而上都有几处空中广场。

该建筑设计使用了与通常办公建筑相反的处理手法（利用建筑周围带有战略性开放的几个体量的开放式交通核心取代通常带有玻璃幕墙的不透明的交通核心），这样做不仅仅是建筑功能方面的需要，同时也出于保护环境方面的考虑，另外也突出了建筑本身的特色。

该建筑设计必须符合客户心目中的"创新中心"建筑看上去一定要"现代"的要求。但盲目地、不加批判地追求当代性使圣地亚哥到处都是玻璃大楼。由于当地炎热的沙漠气候，玻璃大楼内部会形成严重的

温室效应，因此这类塔楼不得不在空气调节方面浪费大量能源。使建筑物避免吸收过多热量不是搞火箭研究，并不是一件多么难的事，只要做到如下三点就足够了：将空间设置在建筑边缘；立面开窗均向后退，避免阳光直射；开窗设计允许空气对流。通过这种做法，我们把建筑的能耗从每年120kW/m^2（这是圣地亚哥常规的玻璃大楼的能耗）降到每年45kW/m^2。创新大楼不透明的外立面不仅十分节能，同时还保护建筑内部免受过强光线的照射。通常，为了保护内部工作空间免受过强光线的照射，人们只能拉上窗帘或百叶窗。这种做法事实上把最初理论上的透明性变成了文字游戏。从这种意义上讲，呼应环境的这种做法不过是对常识加以充分运用罢了。

另一方面，我们认为创新中心面临的最大威胁就是过时，即功能以及风格上的过时。所以，摒弃玻璃外立面的设计不是仅仅是为了避免极端气候环境的专业责任感，也是为了寻找能够经得起时间考验的设计。从功能角度来讲，我们认为避免设计过时的最好方法就是把这栋楼当成基础设施而不是建筑来设计。清晰、直接甚至硬朗的建筑形式证明是最灵活的设计，可以不断变化和更新，具有极强的适应力。从风格角度来讲，我们利用极为严整的几何造型和极为统一的整体性来抵御潮流的变迁，使之成为永恒之作。

Anacleto Angelini UC Innovation Center

In 2011, Angelini Group decided to donate the necessary funds to create a center where companies, businesses and more in general, demand, could converge with researchers and state of the art university knowledge creation. The aim was to contribute to the process of transferring know-how, identifying business opportunities, adding value to existing resources or registering patents in order to improve the country's competitiveness and consequently its development. The Catholic University of Chile would host such a center and allocated a site in its San Joaquin Campus.

Our proposal to accommodate such goals was to design a building in which at least 4 forms of work could be verified: a matrix of formal and informal work crossed by individual and collective ways of encountering people. In addition to that, we thought that face to face contact is unbeatable when one wants to create knowledge, so we multiplied throughout the building the places where people could meet: from the elevator's lobby with a bench where to sit if you happen to run into somebody that has interesting information to share, to a transparent atrium where you can sneak into what others are doing while circulating vertically, to elevated squares throughout the entire height of the building.

The reversal of the typical office space floor plan (replacing the opaque core with transparent curtain wall glass perimeter by an open core with the masses strategically opened in the

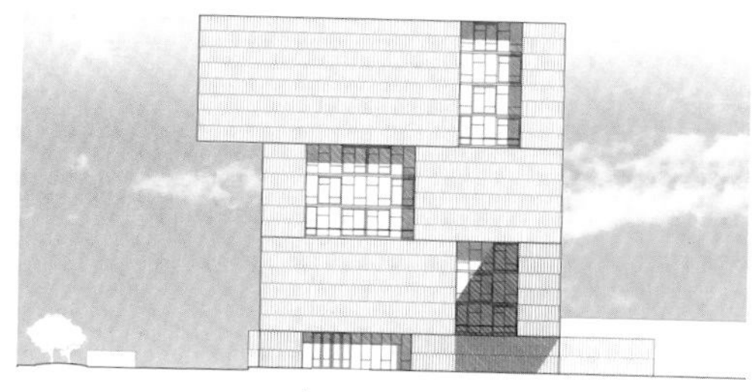

南立面 south elevation

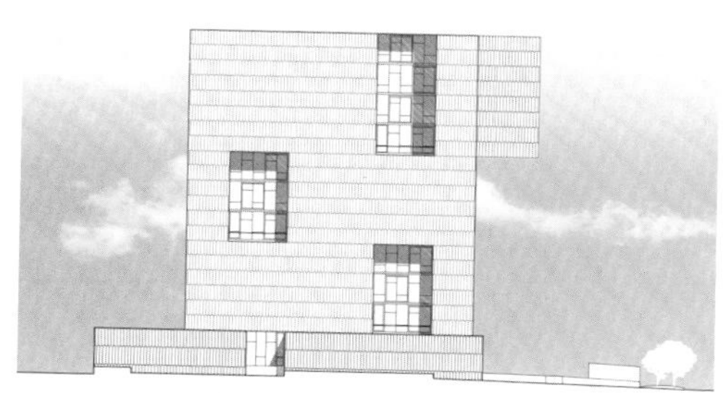

北立面 north elevation

东立面 east elevation

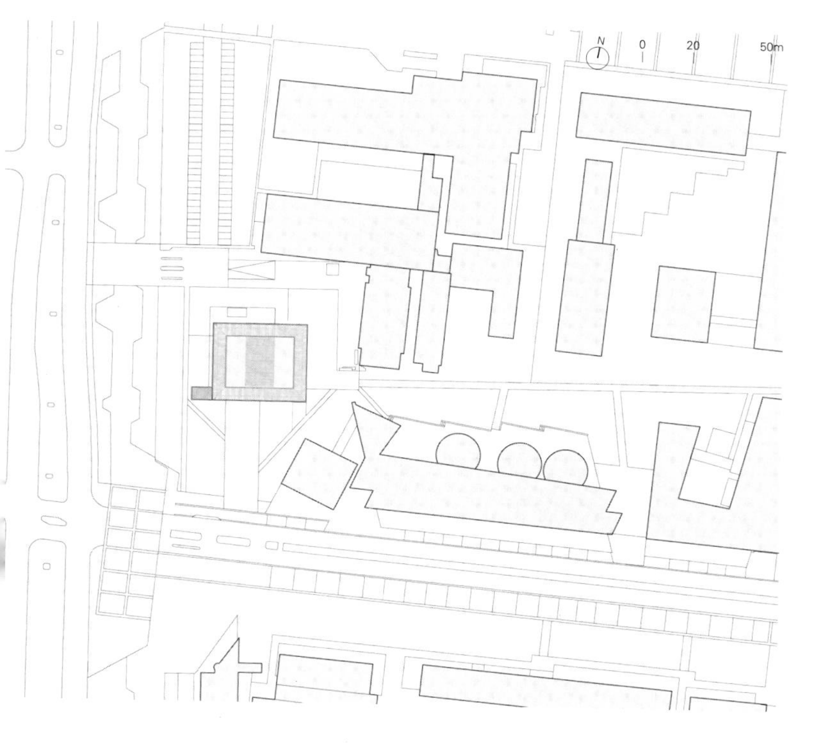

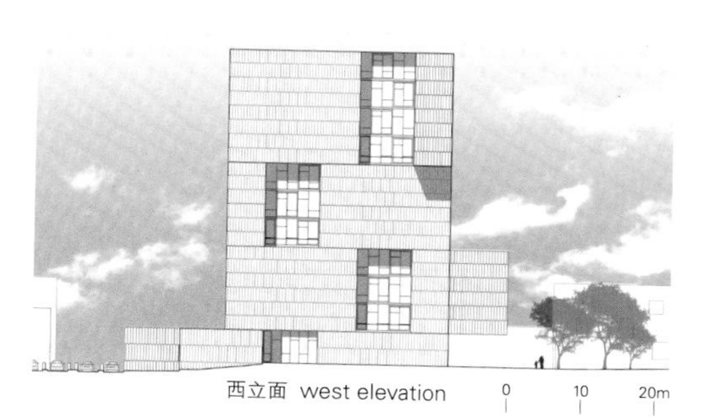

西立面 west elevation 0 10 20m

©James Florio(courtesy of the architect)

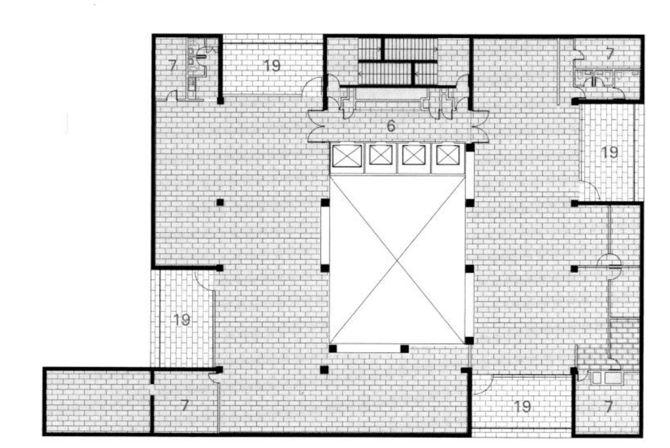

八层 seventh floor

1 主大堂
2 自助餐厅
3 安保室
4 保管室
5 接待处
6 电梯厅
7 餐厅
8 银行办公室
9 储藏室
10 多功能室
11 租用办公区
12 大厅
13 休息室
14 会议室
15 教室
16 视频会议室
17 网络学习室
18 露台
19 抬高的广场
20 合作空间
21 原型制作和电子阅览室
22 办公室
23 研磨区
24 机器人区
25 数据室

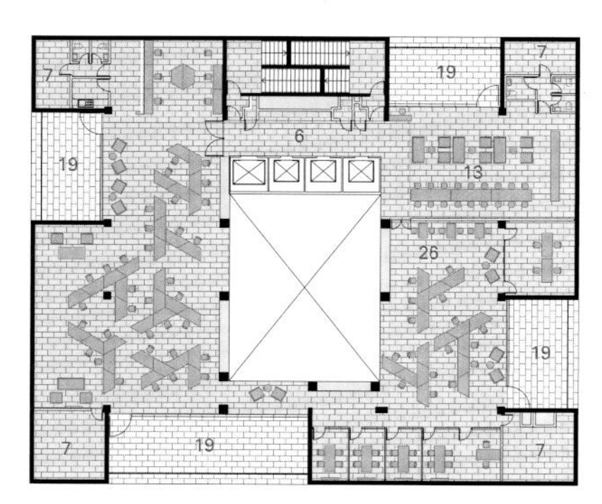

五层 fourth floor

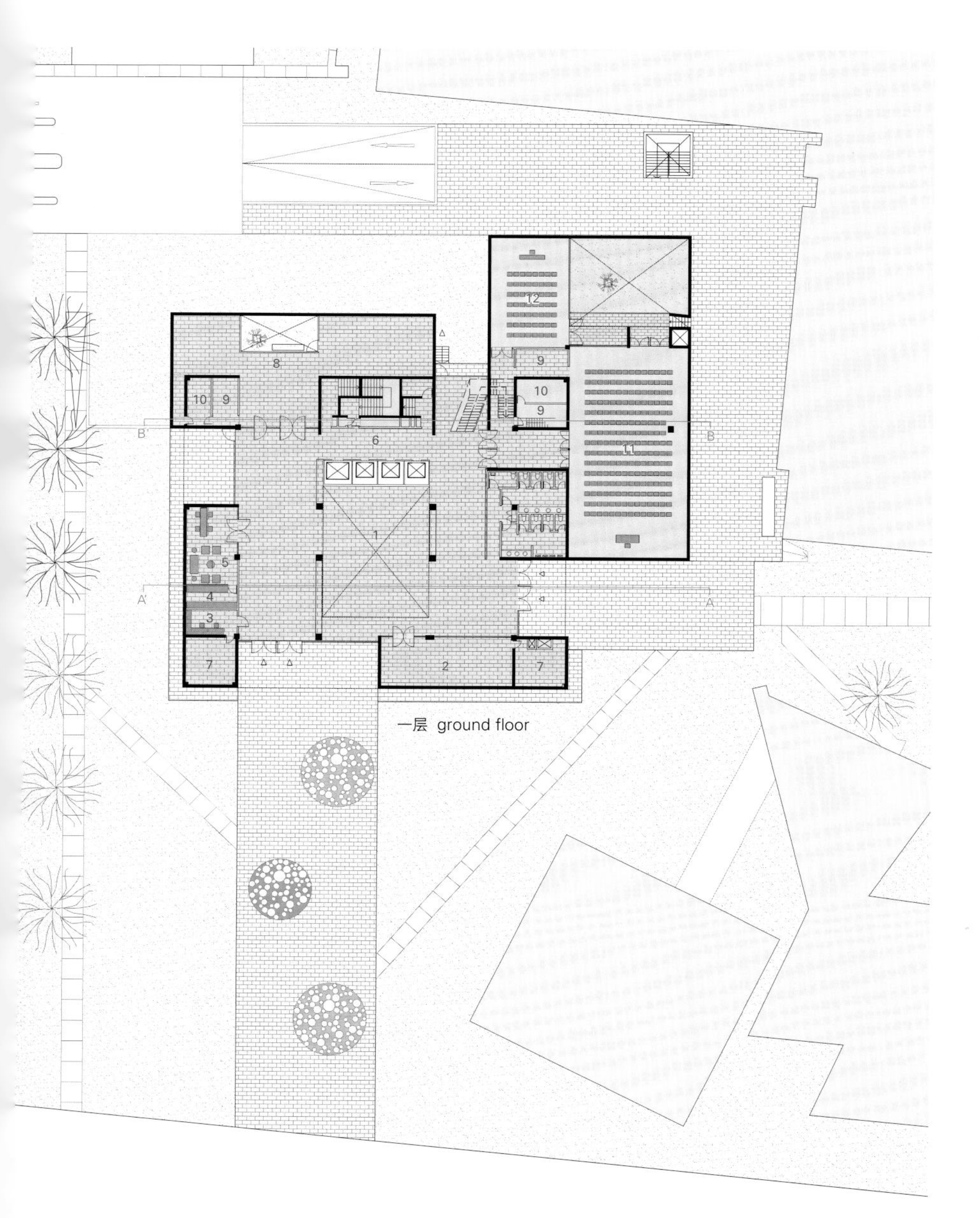

一层 ground floor

三层 second floor

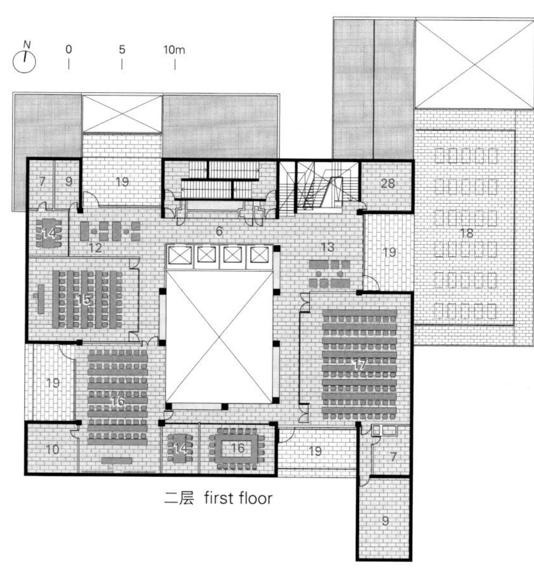

二层 first floor

1. main lobby
2. cafeteria
3. security office
4. safekeeping
5. reception
6. elevator lobby
7. restaurant
8. bank office
9. storage
10. multipurpose room
11. rental office space
12. hall
13. lounge
14. meeting room
15. classroom
16. video conference room
17. e-learning room
18. terrace
19. elevated square
20. coworking space
21. prototyping and electronical area
22. office
23. milling area
24. robot area
25. data room

A–A' 剖面图 section A-A'

B–B' 剖面图 section B-B'

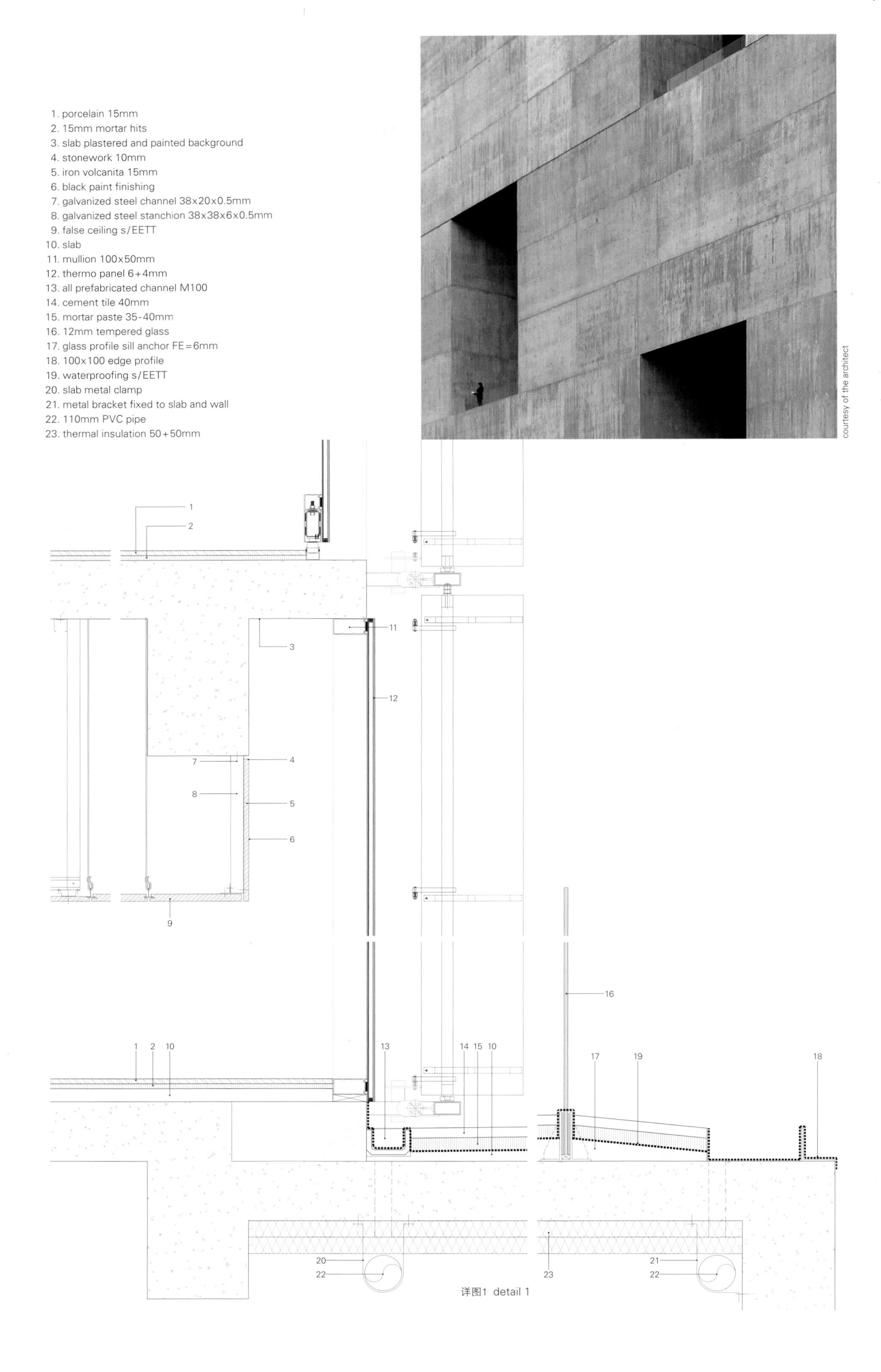

courtesy of the architect

详图1 detail 1

perimeter) responded not only to functional reasons but to the environmental performance and character of the building as well.

This building had to respond to the client's expectation of having an innovation center with a "contemporary look", but the uncritical search for contemporariness has populated Santiago with glass towers that due to the desert climatic local condition have serious greenhouse effect in interiors. Such towers spend a huge amount of energy in air conditioning. The way to avoid undesired heat gains is not rocket science; it is enough to place the mass of the building on the perimeter, have recessed glasses to prevent direct sun radiation and allow for cross ventilation. By doing so we went from 120 kW/m²/year (the consumption of a typical glass tower in Santiago) to 45kW/m²/year. Such an opaque facade was not only energetically efficient but also helped to dim the extremely strong light that normally forces to protect interior working spaces with curtains and blinds transforming in fact, the theoretical initial transparency into a mere rhetoric. In that sense the response to the context was nothing but the rigorous use of common sense.

On the other hand, we thought that the biggest threat to an innovation center is obsolescence: functional and stylistic obsolescence. So the rejection of the glass facade was not only due to the professional responsibility of avoiding an extremely poor environmental performance, but also a search for a design that could stand the test of time. From a functional point of view, we thought the best way to fight obsolescence was to design the building as if it was an infrastructure more than architecture. A clear, direct and even tough form is in the end the most flexible way to allow for continuous change and renewal. From a stylistic point of view, we thought of using a rather strict geometry and strong monolithic materiality as a way to replace trendiness by timelessness. Elemental

← FABLAB

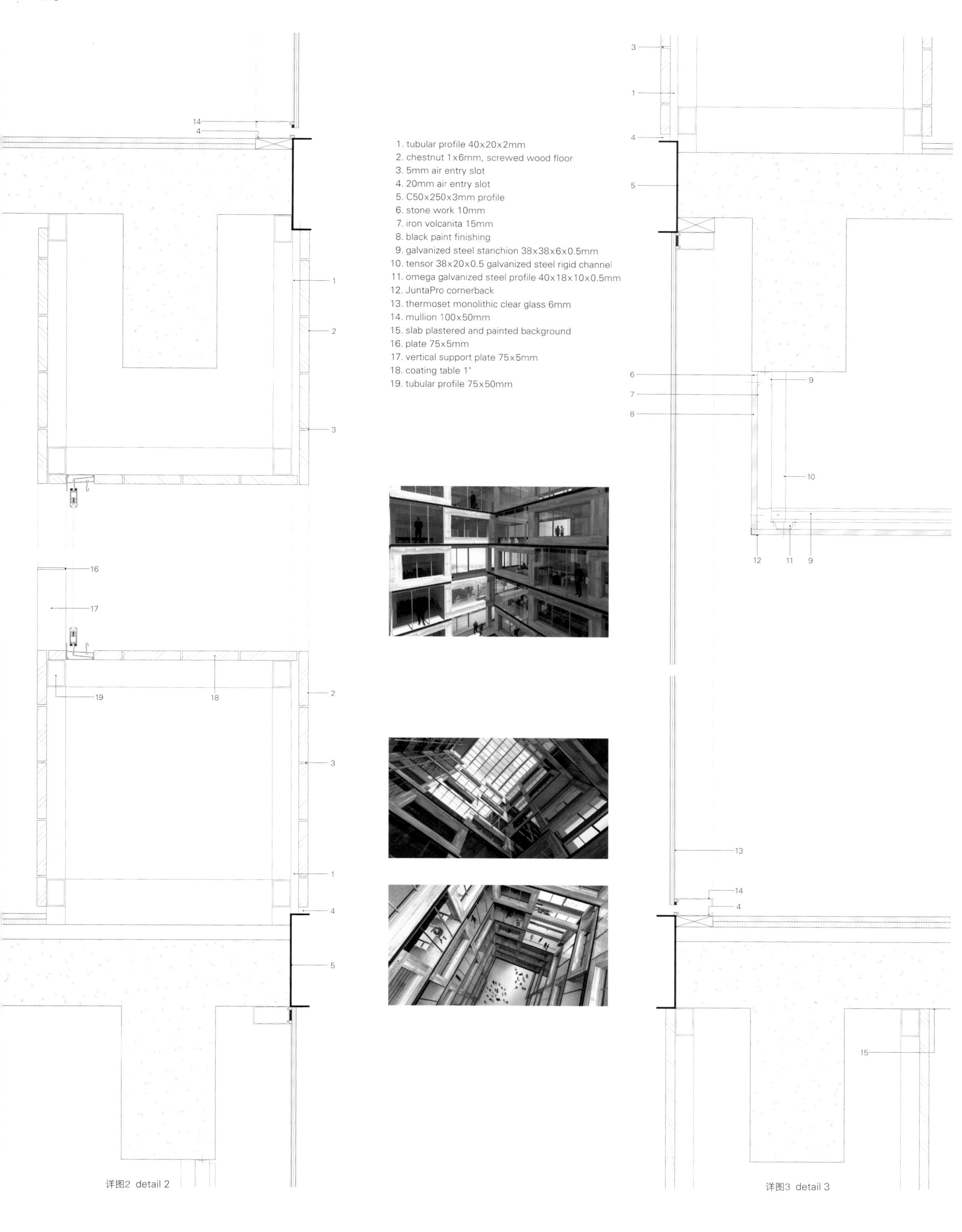

详图2 detail 2

详图3 detail 3

项目名称：Innovation Center UC–Anacleto Angelini
地点：Santiago, Chile
建筑师：Alejandro Aravena, Juan Cerda,
Gonzalo Arteaga, Víctor Oddó, Diego Torres
合作者：Samuel Gonçalves, Cristián Irarrázaval, Álvaro Ascoz
Natalie Ramirez, Christian Lavista, Suyin Chia, Pedro Hoffmann
结构工程师：Mario Alvarez _ Sirve S.A.
电气工程师：Carlos Gana _ Ingenieria y Proyectos ICG y Cía. Ltda
机械工程师：Sirve S.A.
场地监督：Juan Cerda
节能设计：Bustamante y Encina Asesorías en Sustentabilidad
修复：Gerardo Sepúlveda _ S&C Revisores de Edificación
管道工程：Vivanco y Vega Ltda.
空调设计：Gustavo Concha _ A&P Ingeniería
甲方：Grupo Angelini, Pontificia Universidad Católica de Chile
用地面积：455,351m^2
有效楼层面积：8,176m^2 _ building, 12,494m^2 _ parking
材料：reinforced concrete, steel, wood, glass
造价：USD 18 millions
设计时间：2011—2012
施工时间：2012—2014
摄影师：©Felipe Diaz Contardo (except as noted)

户外房间和室内广场
围合社区

Outdoor Rooms
Enveloping Communities

在本文中，我们将探讨一系列小型社区建筑，它们规模不一，是现代建筑的典范，每一个都具有本地社区的功能。在超链接的21世纪，随着虚拟社区和信息服务业的发展，人们也许会说小型社区建筑并没有什么意义，然而这些建筑项目表明人们对当地高质量的城市建筑仍有强劲的需求。每一个项目都致力于为用户既提供像公共广场这类开放的公共使用的空间，又提供一些专门用于某些用途的封闭空间，更加私密，更有家的温馨舒适，并努力在这两者之间架起桥梁。这些设计策略服务于公众的某些民主观念，同时，这些建筑本身也是社区"自我形象"建设的一部分。

In this text we will examine a series of small community buildings, examples of contemporary architecture at a variety of scales, each of which provides a local community function. In the hyper-connected 21st century world one might argue that small community buildings may not have as significant a role to play, with online communities and information services, but these projects show that there is still strong demand for high-quality local civic architecture. Each of these projects is united by an attempt to create spaces that bridge between open democratic uses such as public squares, and the more private and domestic functions of enclosed architecture with specific programmatic uses. These strategies are in the service of a certain democratic notion of the public, and the buildings themselves are part of the construction of their communities' self-image.

nd Indoor Squares

围合社区

在当今世界，信息和媒体都普遍电子化，人们在家庭环境中就可以获得信息和接触各种媒体，这和以往大不相同。城市规划发展的压力在城市的功能性方面有所降低。更多传统的民用市政建筑现在不得不重新评估自己在数字时代的作用，每一个新的社区项目都要考虑人们对公共空间和公共服务需求的不断变化。

最近我们考察了很多建筑项目，大部分是图书馆和一些小型博物馆。这些项目都尽力找到自己作为小型建筑的定位，其所提供的服务如今大多在网上可以获得。在考察其他同一类型的建筑的时候（在本章中考察的是小型社区中心和教育设施），我们也能看到类似的问题和并存的机遇。这些建筑生动地展示了现代建筑设计所采用的一些类似方法，使小型民用市政建筑的设计大胆而活泼，使其在更为传统的住宅和商业建筑中成为意义非凡的地方资产。

关于这类小型民用市政建筑，我们能想到的其他设计方法包括重新设计利用历史建筑，将历史建筑与新的建筑元素融为一体，使建筑与周围环境相得益彰，坚持成为社会发展的有机品质。然而，也许在这些建筑项目中，我们所能看到的最有趣的建筑特色是将公共空间引入建筑。有的建筑中设计了很大的房间用作公共空间，而有的建筑与此恰恰相反，通过定义户外空间，使其就像是建筑物额外的一个房间，把家庭生活的温馨带到了公共领域。

但是这些建筑都不大，不足以为整个城市服务，它们的规模从小型建筑到中型建筑不等。中型规模的建筑一般为小城镇提供服务，而小型建筑如一些活动中心，一般建于村庄或者街道社区。有时候，这些建筑取代了中央露天广场或绿地，然而更多时候则是嵌入社区的剩余空间里。无论哪种情况，其目的都是创建一个将当地人聚集起来的活动中心。

以北京为例，在过去几十年的建设热潮中，许多传统的胡同居民区都消失了，但是在最近几年，人们在努力拯救、修复和保存在一些小规

Enveloping Communities

In a world where information and media are available electronically and in a domestic setting like never before, the pressure in urban development is for a reduction in functions in the city. More traditional civic buildings are having to reevaluate their role in the digital age, and each new community project has to engage with a changing experience of what public space and public services entail.

We recently examined a number of projects, mostly libraries and small museums, that were attempting to find their place as small institutions offering services that today are largely available online. We can see similar problems and opportunities at work in another selection of buildings, in this case small community centers and educational facilities. These buildings exemplify similar approaches to designing contemporary institutions, where a small civic building is given an opportunity to be bold and outgoing, to mark it as a significant local asset amongst more traditional residential and commercial buildings.

Other strategies that we can discern include the re-use and incorporation of historic buildings into new developments, tying them into the fabric of their locations, asserting the organic quality of their social development. But perhaps the most interesting architectural quality we see in these projects is the bringing of public space into the building, through creating large rooms mimicking public space, or the opposite, the definition of outdoor spaces that become something akin to an extra room of the building, an introduction of a certain domesticity to the public realm.

While none of these buildings is large enough to serve entire cities, they range in scale from medium projects that offer services to small towns, to very small buildings that create centers within villages and neighbourhoods. In some cases they replace a central open plaza or green space, while in others they are inserted into more residual spaces within a community. In both cases they are intended to create a focus

线图提供：©Hascher Jehle Architektur

微院儿，北京，中国
Micro-Yuan'er in Beijing, China

模、高密度居住的具有代表性的胡同建筑。为了"突出这些附加建筑重要的历史地位，Standardarchitecture建筑事务所设计了"微院儿"。"微院儿"是一处微型儿童图书馆和艺术空间，依树而建。图书馆镶嵌在原有建筑的壳里，室内由简单的木材搭建而成，屋顶低垂，就像树的枝干。艺术空间的外立面由当地的灰砖垒成，与周围环境保持一致。一段楼梯依树而建，直达屋顶，为孩子们营造了一处小小的空间，孩子们可以坐在屋顶之上，树枝之间。艺术空间看起来好像与图书馆完全分割开来，暗示了在原来开放的露天空间中形成的另一个更小的庭院。

有一个叫伊施格尔的小村庄依偎着奥地利阿尔卑斯山。在这里，Parc Architekten建筑事务所为设计位于村庄中央的文化中心几乎使用了相同的建筑策略。这个文化中心的主要用途是为当地的音乐俱乐部提供排练空间，但也供举行其他社会活动使用。文化中心的位置以前曾经是村中心的一块绿地，所以其大部分建筑嵌在这个绿草茵茵的陡坡内，露出地表的部分就像草地上露出的岩层。

这栋建筑还有其他的设施。电梯向下与地下通道相连，供滑雪者使用；同时它还与村子里的一座历史建筑相连接。通过精心设计，该建筑物前面有一处室内开放空间，其中一端是覆顶的舞台，另一端是台阶，台阶作为一座露天礼堂使用。滑雪季节是这个村子最忙的时候，这里就不太可能派上多少用场。当天气暖和的时候，也是一年中村子相对来说比较安静的时候，这里就成为开放的社区活动场所了。

费尔德基希是位于阿尔卑斯山脚下的另外一座奥地利小镇，它以保存完好的中世纪小镇中心区而闻名。Hascher Jehle建筑事务所和Mitiska Wäger建筑事务所共同为小镇设计建造了一个非常现代的新文化中心，所采用的建筑设计策略与伊施格尔镇文化中心的相似，但是规模更大。

这个文化中心的建筑布局以一个大型菱形礼堂为中心，走廊楼梯等流通空间和其他服务设施都围绕礼堂而设计。玻璃屋顶下是三层高的交通流通空间，许多颇为壮观的楼梯曲折蜿蜒，与礼堂外部紧密相

of activity to bring local people together.

In Beijing, for example, many traditional hutong neighbourhoods have disappeared in the building boom of the last decades, but in recent years attempts have been made to rescue, renovate and save a few small examples of the these small-scale, high-density residential zones. In order to "recognise the add-on structures as an important historical layer" of Beijing, Standardarchitecture built Micro-Yuan'er, a tiny children's library and art space, wrapped around a tree.

The library is a simple timber interior, with a ceiling that bows like the branches of a tree, inserted inside the shell of an existing building. The art space is faced with local grey bricks, mimicking the appearance of the surroundings, while around the tree a staircase leads up to the roof, creating a small space where children can sit amongst the branches. The second structure seems almost as though it has been pulled apart from the first, implying a second, smaller courtyard within the original open space.

Nestled into the Austrian Alps is a small village called Ischgl, where Parc Architekten have deployed a rather similar architectural strategy for a culture center in the middle of this small village. The primary use for the building is rehearsal space for the local music group, but it also has rooms for other social activities too. Embedded into a steep slope, on land that had previously been a green space at the village center, most of the accommodation is underground, breaking out of the grassy landscape like rock strata.

Other facilities provided by the building are a lift that connects down to a tunnel for use by skiers, as well as a connection to one of the historic buildings of the village. The building uses its plan to create a sheltered open space in front, with a covered stage at one end, and steps that function as an auditorium at the other. While this is unlikely to be much use during the skiing season when the village is busiest, it provides an open community space during the warmer, quieter times of year.

Feldkirch is another Austrian town at the foothills of the Alps, known for its well-preserved medieval central area. Into this zone, Hascher Jehle Architektur and Mitiska Wäger Architekten have built a new cultural center to a very contemporary

照片提供：©David Schreyer

圣尼古拉斯文化中心，伊施格尔，奥地利
St. Nikolaus Cultural Center in Ischgl, Austria

连。建筑师解释说："室外空间到室内空间的这种无缝流动连接，以及建筑大厅整面玻璃幕墙的设计，使周边环境成为建筑空间的一部分，并营造了内部空间的空旷感。"

费尔德基希的Monforthaus文化中心的设计在平面、剖面和整个立面方面都大量使用曲形表面来营造这种空间的"流动感"。这一设计在很大程度上借鉴了扎哈·哈迪德建筑师事务所提出的"参数化"设计方法，在规模和建筑材料的选择方面，都想方设法地与当地城市的整体效果融为一体。

另外一个文化中心位于法国南部的小镇圣热尔曼阿尔帕容，由O-S建筑师工作室设计。文化中心由一座音乐舞蹈学校和图书馆组成。乍一看，它与伊施格尔和费尔德基希的文化中心完全不一样，其外立面简朴无趣，由一系列整齐划一的青铜色铝面板装饰，从外观看是非常传统的城市建筑。但是，它们都采用了很多同样的建筑设计原则和理念。

这个文化中心同样也坐落在一个陡坡上。其两个主入口用途不一。一个从上面可以进入图书馆，直达一个大型玻璃幕墙，从这里可以俯瞰周边环境。建筑物的两翼之间有一段陡峭的楼梯，沿着楼梯向下可以到达音乐舞蹈学校的入口，之后沿着楼梯向下可以到达建筑物的另一侧。同样，这种建筑策略形成了一处安全的内部房间：一座庭院。这座庭院也可用作走廊。

法国北部的Madeleine媒体图书馆由Tank建筑师事务所负责设计，主要是对以前用作警察局的这一新古典主义时期建筑进行扩建。Madeleine媒体图书馆的大部分设施都位于带有传统布局的原来建筑内，但是新建筑物的结构设计主要通过屋顶设计来定义。屋顶根据几何学设计布局，三角形组成的屋顶图案就像盛开的花朵，且屋顶照明与平面布局的深度相呼应。

这处空间几乎完全是一间单一的没有分割的房间。建筑师称这一空间"成为沿街广场的延伸"。因此，Madeleine媒体图书馆不是一座自

design, one that employs a similar architectural strategy to the Ischgl Building but on a much larger scale.

The organisation of the building is gathered around a large, lozenge-shaped auditorium, with circulation and service spaces arranged around it. Adjacent to this is a triple-height circulation space under a glazed roof, with a number of grand staircase curving around the exterior. The architects explain that "the seamless flow of space from outside to inside and the fully glazed walls of the foyer make the surroundings part of the space and contribute to the sense of an expansive interior."

Montforthaus in Feldkirch expresses this "flow" of space with its frequent use of curved surfaces, in plan, section and across the elevations. Owing much to the "parametricist" approach to architecture developed by Zaha Hadid Architects, it nevertheless attempts to become part of the local urban ensemble through the scale and mimicry of material.

Another Cultural Center, designed by Atelier O-S Architectes at Saint-Germain-lès-Arpajon, a town to the south of Paris, incorporates a music and dance school, as well as a library. At first sight it appears completely different to the Ischgl and Feldkirch Buildings, with its austere facade made up of a series of regular bronze coloured aluminium panels, and its more conventional urban appearance, but they share many of the same principles.

The Cultural Center is also embedded into a sloping site. Two main entrances are provided for the different uses, with the library entered from the top of the site, leading to a large glazed facade looking out over the surroundings. A steep set of stairs leads down between two wings of the building to the music and dance school entrance, and then leads underneath the building to the other side. Again, this strategy creates a safely internal room, a courtyard that also functions as a corridor.

Madeleine Media Library, in northern France, was designed by Tank Architectes as an extension to a Neoclassical former police station. Most of the facilities are gathered into the older building with its conventional layout, but the new building structure is primarily defined by its roof, which is geometrically arranged in an almost flower-like triangulated pattern, with

线图提供：©Tank Architectes

Andree Chedid媒体图书馆，图尔宽，北部，法国
Andrée Chedid Media Library in Tourcoing, North, France

我封闭、为居民提供一间户外房间的社区建筑，而是将传统意义上开放的公共空间延伸至室内，将这一设施尽可能地开放和为公众所用。

人们从Madeleine媒体图书馆驱车不到30分钟，就可以看到另外一座图书馆建筑。由D'HOUNDT+BAJART建筑事务所设计的这座图书馆几乎采用完全一样的建筑方法，但其结构更加大胆和冒险。这座图书馆也是利用一处历史遗址修建的。占据这一地块的小型建筑曾经是一个纺织工厂的一部分。

这座小型历史建筑是图书馆的"核心"空间，有卫生间和其他一些设施，宽阔的一层用作礼堂等其他图书馆设施。大型屋顶并没有采用圆柱支撑，而是用木材搭建成了蜂窝形。这种建筑方法再次在建筑物内营造了开放式景观的效果。不仅建筑外立面用六边形建筑元素装饰，整座建筑物都使用与之呼应的六边形建筑元素，包括家具、装饰以及其他元素。

维谢格拉德镇是匈牙利一座非常小的城镇，位于布达佩斯北部。2000年，当aplusarchitects事务所第一次开始为维谢格拉德镇设计新的镇中心时，他们的设计蓝图非常宏大并且全面，包括市政厅、文化中心和许多其他的公共设施。经济危机虽没有终止这个项目的建设，但是其规模却大大降低，几乎完全不考虑建造新的建筑。

活动大厅是最终所完成的这个项目的主要部分，也是一个全新的结构。与前面所提到的刻意将现代设计与历史建筑结合在一起的设计策略相比，这个活动大厅的设计几乎完全依靠周边建筑和现有建筑结构。活动大厅为狭长的木质建筑，带有斜屋顶，厚重的外立面的凹进处可以折叠，使活动大厅成为一处几乎完全开放的空间。

活动大厅用来举行地方活动，在旅游旺季也可以举行更大规模的集会，它可以用作户外的房间，也可以是一处封闭的公共空间。建筑师把它描述为"使整处风景锦上添花"。而对面健康水疗馆的整修翻新工作则包括给楼梯穿上了一件木条制成的外衣。这样不仅可以与当地传统建筑相呼应，而且也与小广场那边的新建筑产生共鸣。

rooflights above to counter the depth of the plan.
This space is almost completely a single undivided room, the architects claiming that it "becomes an extension of the square at street level." So instead of a community building wrapping around itself to provide an outdoor room, the Mediateque la Madeleine extends a traditionally open public space indoors, an attempt to render the facility as symbolically open and public as it can possibly be.
Less than a 30 minute drive from La Madeleine, another library building by D'HOUNDT+BAJART architectes&associés engages in a very similar architectural approach, with even more structural adventurousness. The library utilises another historic remnant, in this case a small building that was once part of a textile factory that dominated the site.
The small historic building also provides "core" space, such as WCs and other programmes, with the expansive ground floor given over to the library facilities, including an auditorium. Instead of a large roof spreading out off columns, the structural system is a honeycomb of timber units, which again create a form of open landscape within the building. The hexagonal image from the structure, expressed on the facade, is then repeated throughout the building, echoed in furniture, decoration and other elements.
In 2000, when aplusarchitects first began to design a new town center for Visegrád, a very small Hungarian town north of Budapest, their brief was large and comprehensive, involving a town hall, a cultural center, and a number of public spaces. The financial crisis didn't stop the project, but it reduced the scope drastically, almost eliminating the need for new buildings.
The main part of the project that was eventually built as an original structure is an events hall, and compared to the strategy of attaching a deliberately modern design onto a historic building, this hall is designed to rely almost entirely on the architecture of the surrounding and existing structures. It is a long timber building with a pitched roof, with a thick facade whose bays can be folded back to create an almost completely open space.
This events hall, which can be used for local events or the larger gatherings that occur during peak tourist periods, is

照片提供：©Brown Meneses Arquitectos

巴尼奥斯镇社区中心，厄瓜多尔
Baños Community Center in Baños, Ecuador

就像给旧建筑加了一个新山墙，有时候一个非常小的动作就能使一处公共空间变得更加完整。在厄瓜多尔的巴尼奥斯有一个建于18世纪晚期的教堂，在1949年的地震中被摧毁，从那以后完全被遗弃，并且被后建的建筑所包围。Brown Meneses建筑事务所接受委托，不仅要保护这些废墟，而且还要把它作为一个社区中心重新恢复使用。

建筑师的解决方案既优雅又简洁。首先对已存的遗址墙面加以清洁，在必要的地方进行修复。然后，在风吹日晒的遗址围墙顶部搭建一个木结构。这个木结构承载着教堂大厅上面的聚碳酸酯材料的屋顶。半透明的光线透过屋顶倾泻而下照进大厅。

屋顶结构虽然简单，但是改变木构件的高度且富有动感的双曲面屋顶设计可实现屋顶的丰富性。双曲面屋顶悬于教堂墙壁边缘之上，这明显不是原建筑的一部分，但是作为全新的、轻盈而简洁的扩建部分，它与原建筑并列组合，使原建筑恢复生机，重新恢复使用。

这种轻盈明亮的屋顶设计巧妙地突出了上文所述的所有建筑的特点。所有这些建筑项目都似乎用文字语言清晰表达了所展示的外围护结构特性。他们所服务的社区之间的关系和纽带都是通过建筑形式来实现的，通常这些建筑都位于社区的中心位置。每一个项目都表明，即使建筑物的内容——媒体、信息、教育——在网上都可以找到，但是，人们仍然对可以表达社区的接近和聚集的渴望的建筑有强烈的需求；这些项目都表明希望通过自己的建筑语言来帮助社区不断完善其本身。

both an outdoor room and an enclosed public space. The architects describe it as "an object-like gesture of the complete landscape", and their refurbishment of the facing health spa building involves adding a stair clad in timber ribs that not only connect back to local traditional construction, but also set up a resonance across the square towards the new structure.

Like placing a new gable end onto an older building, sometimes it is a very small gesture that can complete a public space. In Baños, Ecuador, a late-18th century church, destroyed by an earthquake in 1949, had been completely abandoned, and hemmed in by later constructions. Brown Meneses Arquitectos were commissioned not only to protect the ruins, but also bring them back to use as a community center.

The solution of the architects was elegant, and simple. The existing walls were cleaned and repaired where necessary, and then a timber structure was built off the top of the exposed walls. This timber structure holds a polycarbonate roof over the hall of the church, which casts translucent light down into the space below.

The roof structure is simple, but is given richness through varying the height of the timber members, creating a sweeping hyperboloid surface that cantilevers over the edges of the church walls. The roof is clearly not a part of the original building, but juxtaposes itself as a new, lightweight and simple addition that brings the original back into life and use.

The lightness of gesture in this building deftly highlights qualities that all of the previous projects share. It makes almost literal the enveloping qualities that they exhibit, where the relationships and ties of the communities that they serve are given architectural form, often at the physical heart of the community. Each one of these projects shows that even if the contents of the buildings – media, information, education – are all increasingly available online, in solitude, there still remains a strong need for architecture that can express a community's desire for proximity and togetherness, and that in its architectural expression it assist the community in its continual process of creating itself. Douglas Murphy

线图提供©Tank Architectes, D'HOUNDT+BAJART architectes & associés, ZAO / Standardarchitecture

市政府决定改造以前用作警察局的这栋建筑。由选出的代表和Madeleine的居民组成的项目委员会决定在市场的这个地方建一座媒体图书馆，一面是热闹的人群，一面是中产阶级居住的区域。这项工程将图书馆对全市的全面开放性与营造温馨舒适的家的氛围结合起来。

图书馆被设计为家和工作场所之间的"第三个地方"。对原建筑物的处理是呈大面积开放的状态，以变得更通透，捕捉更多的公共空间。地面也进行了改造，阅览室和公共空间区域在一个水平面上，使一层成为一个连续的空间。

原建筑物面朝市场，媒体图书馆的入口和公共空间都在此处，这里有大厅、礼堂、展览厅、咖啡馆和教学法工作坊。接待区很自然地把人们带到了另外一个更加安静的世界：扩展区。开有天窗的木质天花板似乎漂浮在地面上。这处空间是主阅览室，面朝天空、邻近的街道和花园。其富于变化的屋顶将空间或延伸或压缩地分割，提供了更为私人的空间，营造出一种宁静祥和的气氛，吸引着人们接触各种各样的媒体。

媒体图书馆几乎占据了整片地。九个三角形棚状屋顶由一系列非常规则的圆柱体支撑，屋顶上面的玻璃向北开，这样反射进室内的光就会非常柔和，四散开来。屋顶设计与周围环境相得益彰，面向城市景观和花园开放。屋顶的低洼处平缓向下，引导水流通过三个滴水嘴排到花园中。这处完全开放、灵活的空间里共使用了90种木材饰面来营造多样化的空间。

Madeleine Media Library

The Municipality decided to rehabilitate this building which was a former police station. A participative committee composed by elected representatives and inhabitants of La Madeleine decided to implant the media library on the market place between a popular area and a middle-class district. This project combines the wide opening of the library program on the city with the search for a warm and homely atmosphere. The facility has been designed as "a third place" between home and workplace. The existing building is largely open in order to make it more transparent and to capture the public space. The floors have been modified to bring the reference

Mandeleine媒体图书馆

Tank Architectes

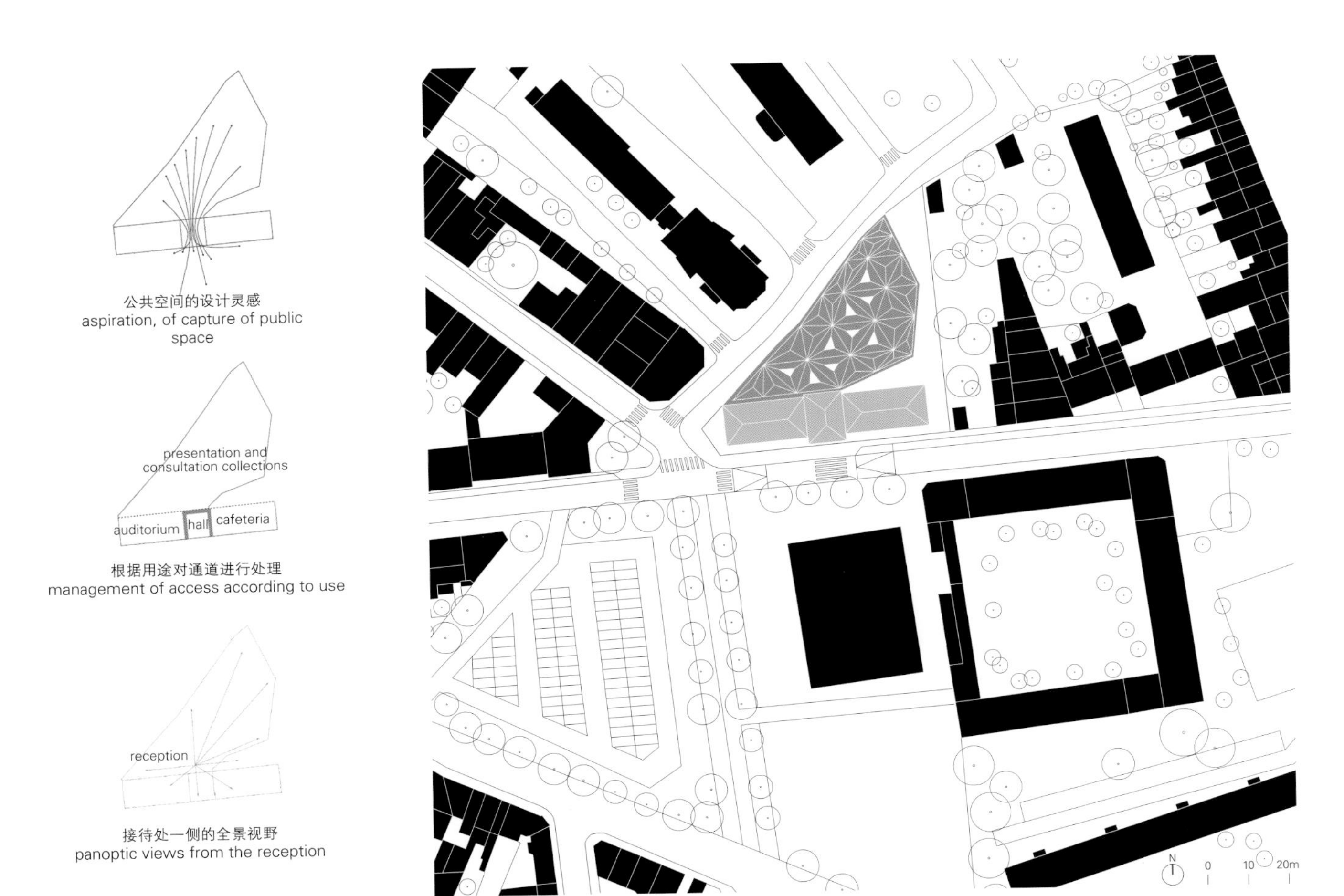

公共空间的设计灵感
aspiration, of capture of public space

根据用途对通道进行处理
management of access according to use

接待处一侧的全景视野
panoptic views from the reception

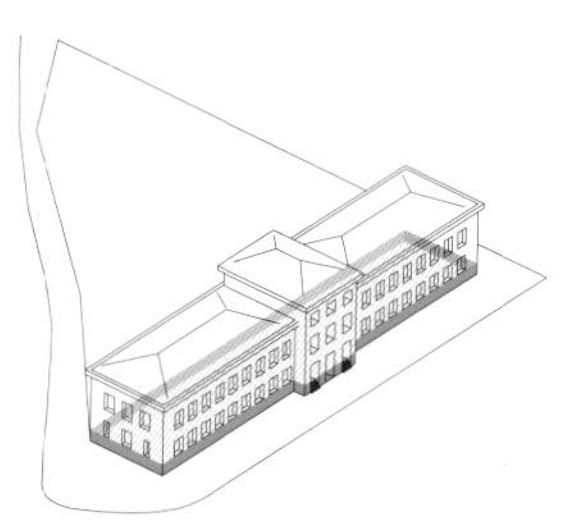

将建筑的基础移至场地的媒体图书馆
removing the base to
a media library at the site

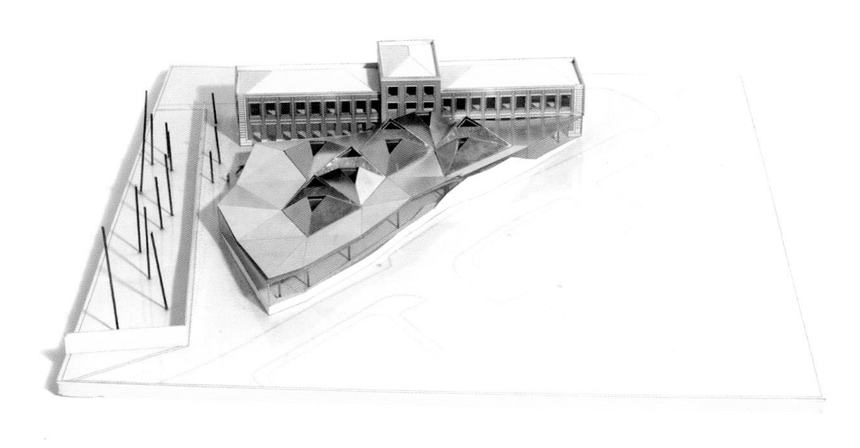

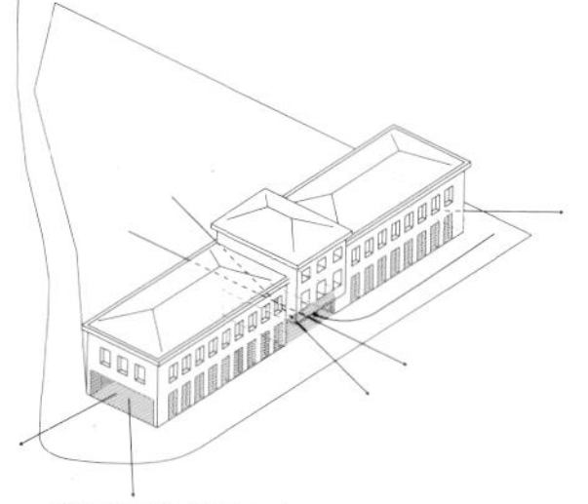

将现有的建筑面向公共空间开放
opening the existing building
to the public space

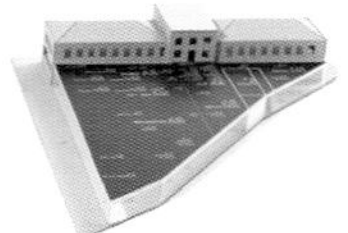

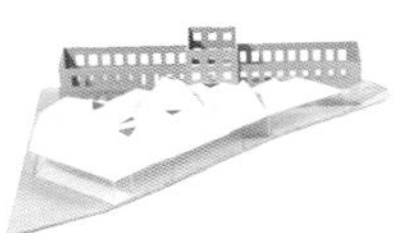

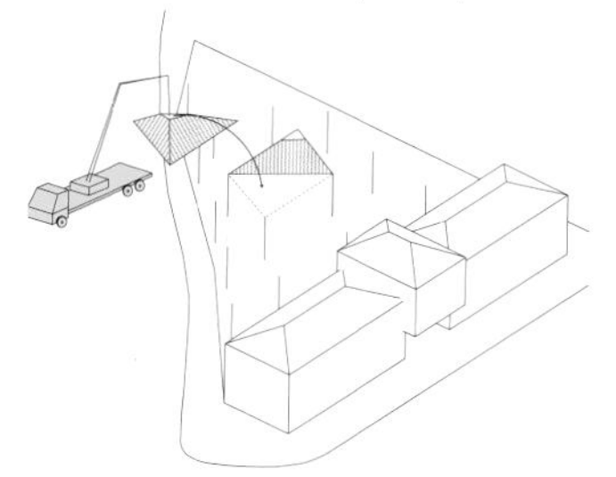

预制盒状结构的屋顶
prefabricating the cover of the boxes

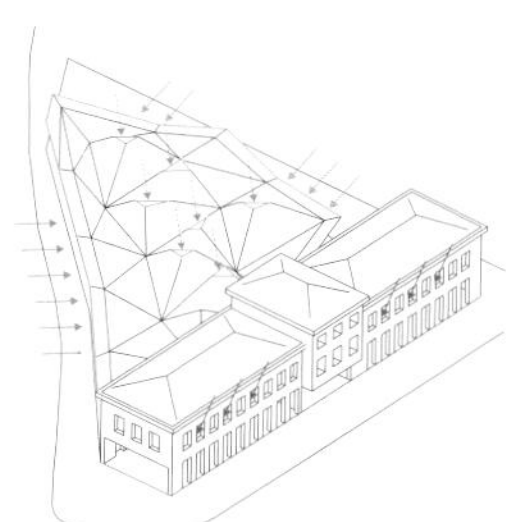

使用自然光来使各个房间变得不同
using natural light
to differentiate spaces

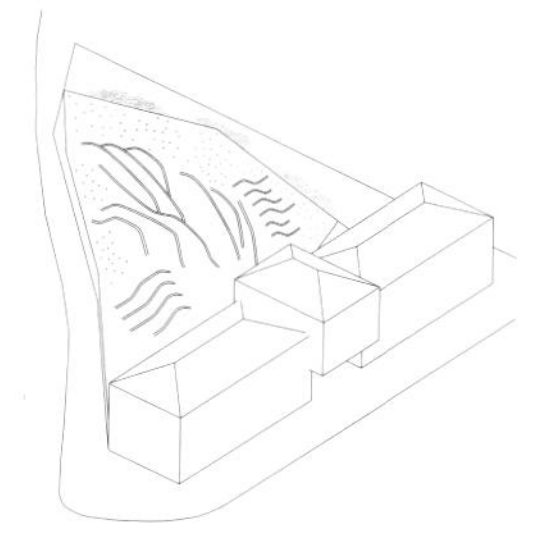

利用家具来建造景观
creating a landscape with the furniture

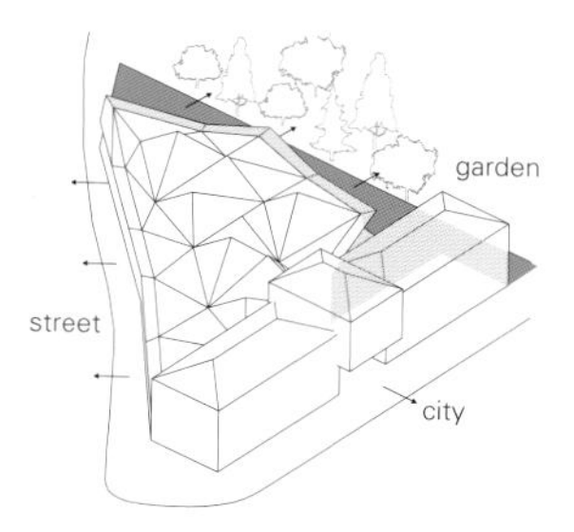

建筑三个侧面进行开放，使其面向花园、城市和街道
opening on 3 sides: garden, city, street

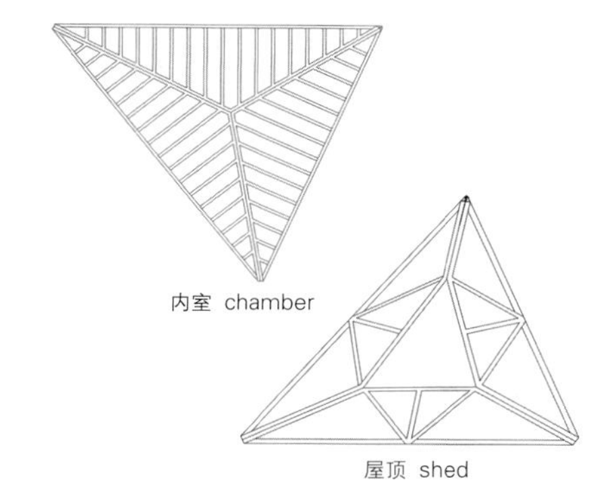

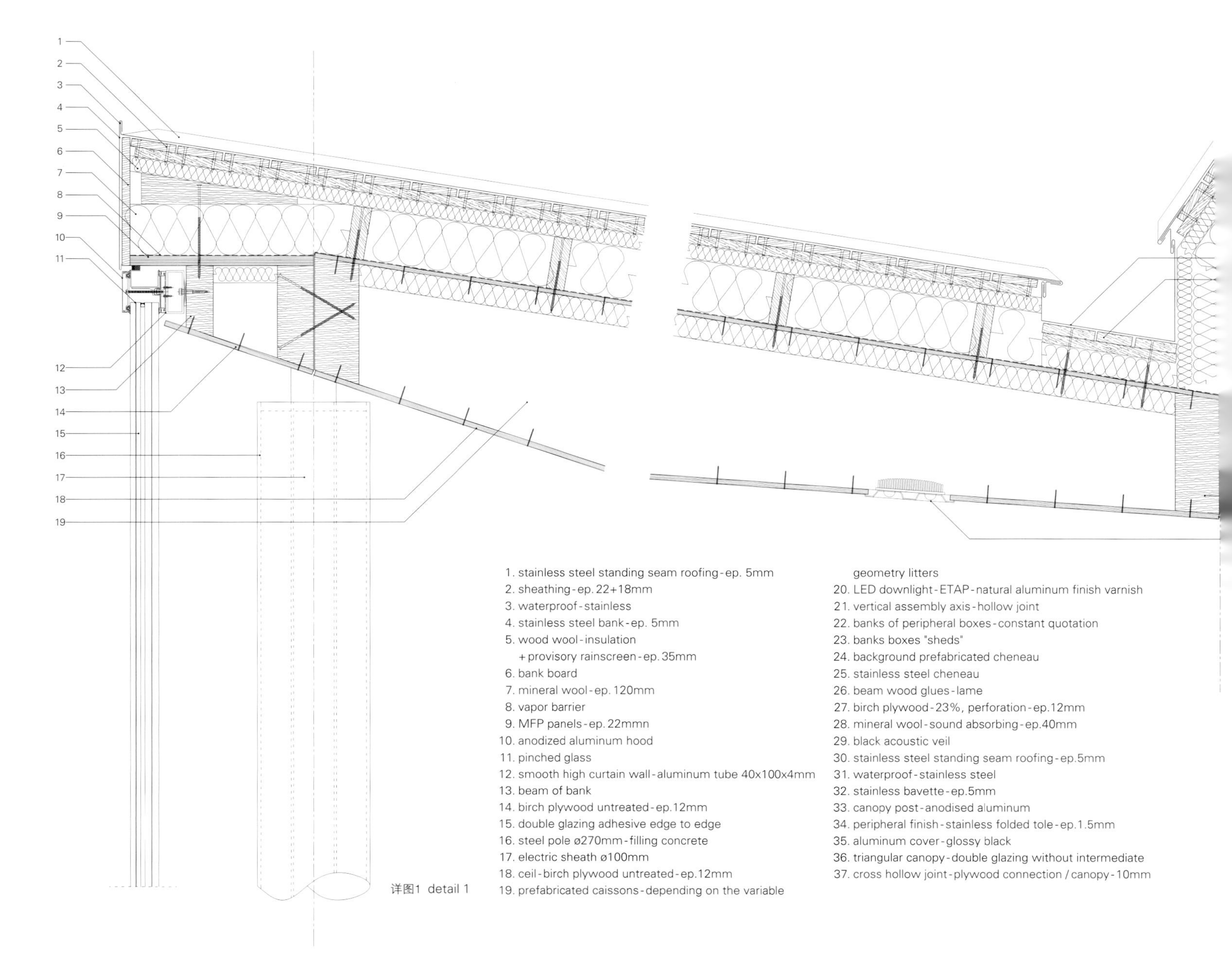

1. stainless steel standing seam roofing-ep. 5mm
2. sheathing-ep. 22+18mm
3. waterproof-stainless
4. stainless steel bank-ep. 5mm
5. wood wool-insulation
   +provisory rainscreen-ep. 35mm
6. bank board
7. mineral wool-ep. 120mm
8. vapor barrier
9. MFP panels-ep. 22mmn
10. anodized aluminum hood
11. pinched glass
12. smooth high curtain wall-aluminum tube 40x100x4mm
13. beam of bank
14. birch plywood untreated-ep.12mm
15. double glazing adhesive edge to edge
16. steel pole ø270mm-filling concrete
17. electric sheath ø100mm
18. ceil-birch plywood untreated-ep.12mm
19. prefabricated caissons-depending on the variable geometry litters
20. LED downlight-ETAP-natural aluminum finish varnish
21. vertical assembly axis-hollow joint
22. banks of peripheral boxes-constant quotation
23. banks boxes "sheds"
24. background prefabricated cheneau
25. stainless steel cheneau
26. beam wood glues-lame
27. birch plywood-23%, perforation-ep.12mm
28. mineral wool-sound absorbing-ep.40mm
29. black acoustic veil
30. stainless steel standing seam roofing-ep.5mm
31. waterproof-stainless steel
32. stainless bavette-ep.5mm
33. canopy post-anodised aluminum
34. peripheral finish-stainless folded tole-ep.1.5mm
35. aluminum cover-glossy black
36. triangular canopy-double glazing without intermediate
37. cross hollow joint-plywood connection / canopy-10mm

详图1 detail 1

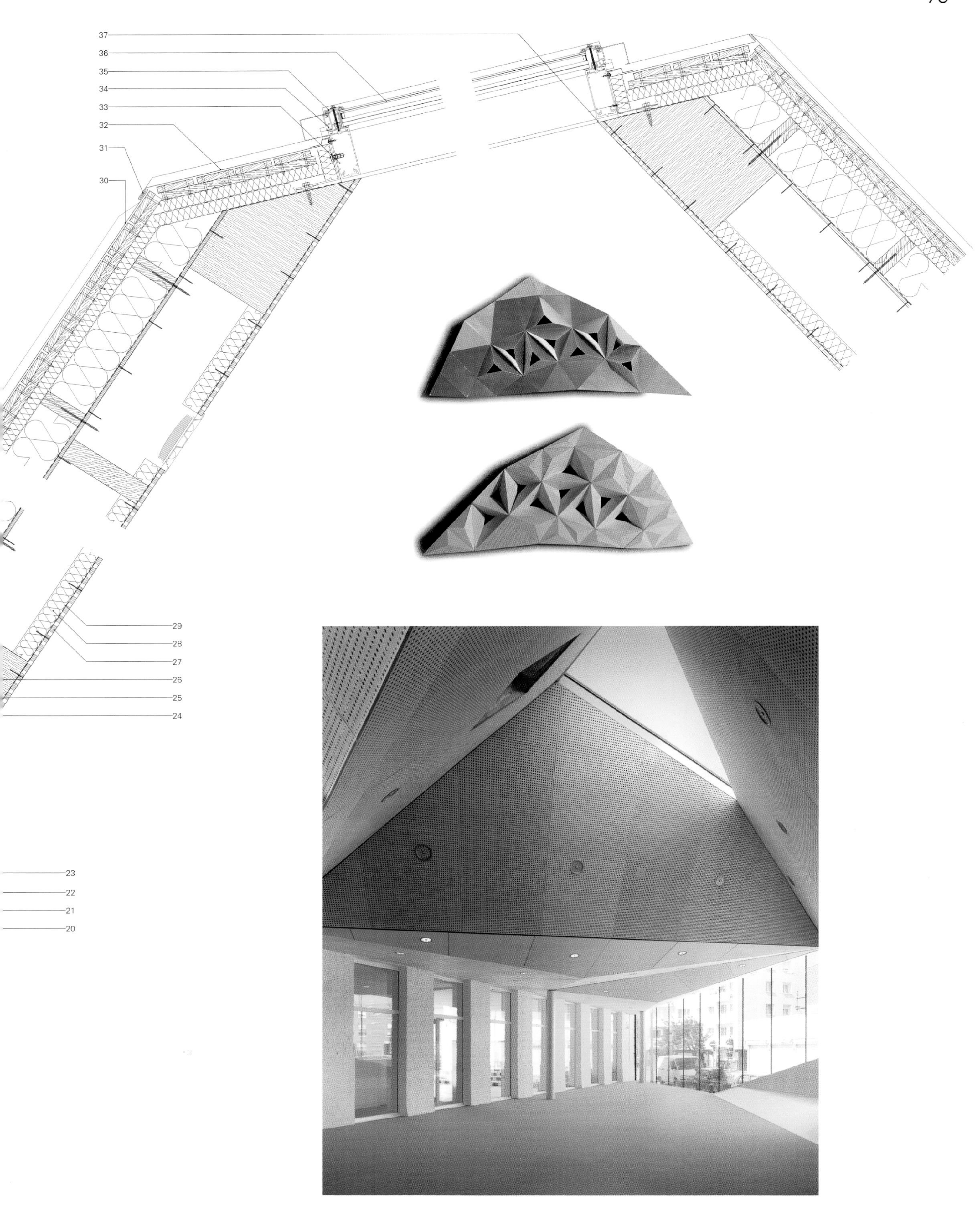
37
36
35
34
33
32
31
30
29
28
27
26
25
24
23
22
21
20

UV
W
WXYZ
Albums tout-petits

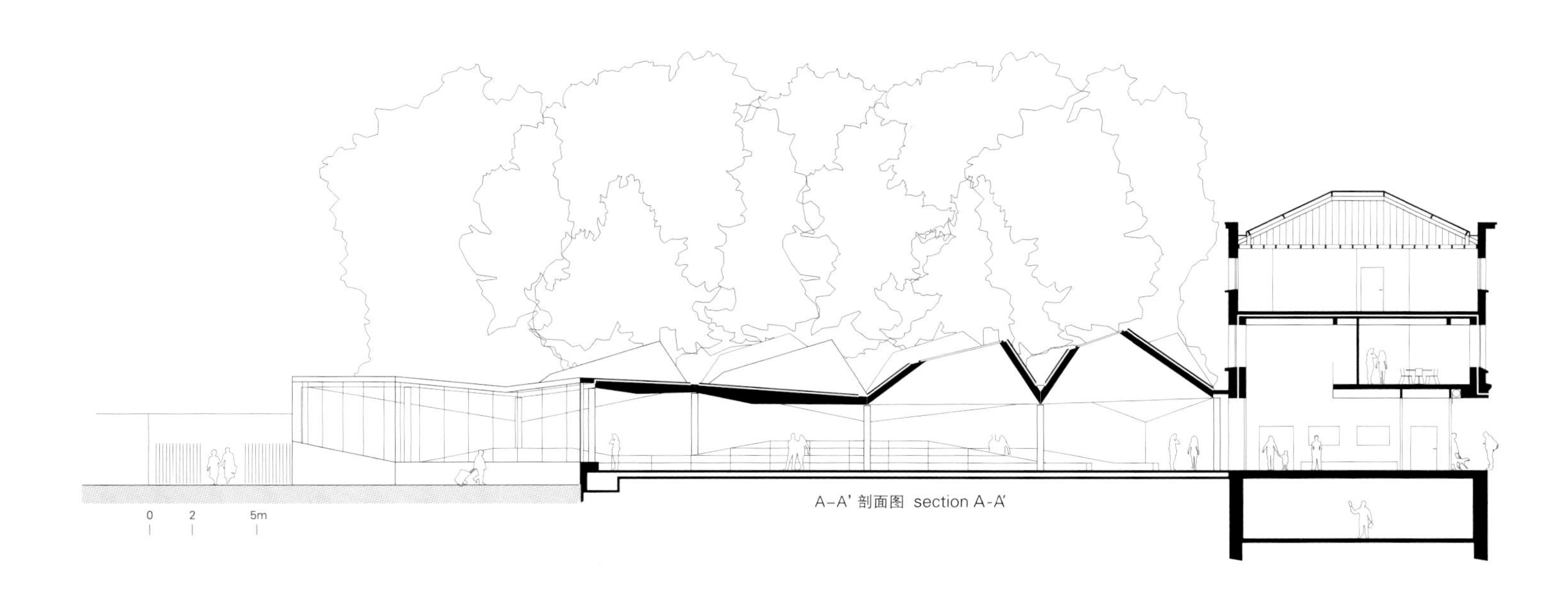

A–A' 剖面图 section A-A'

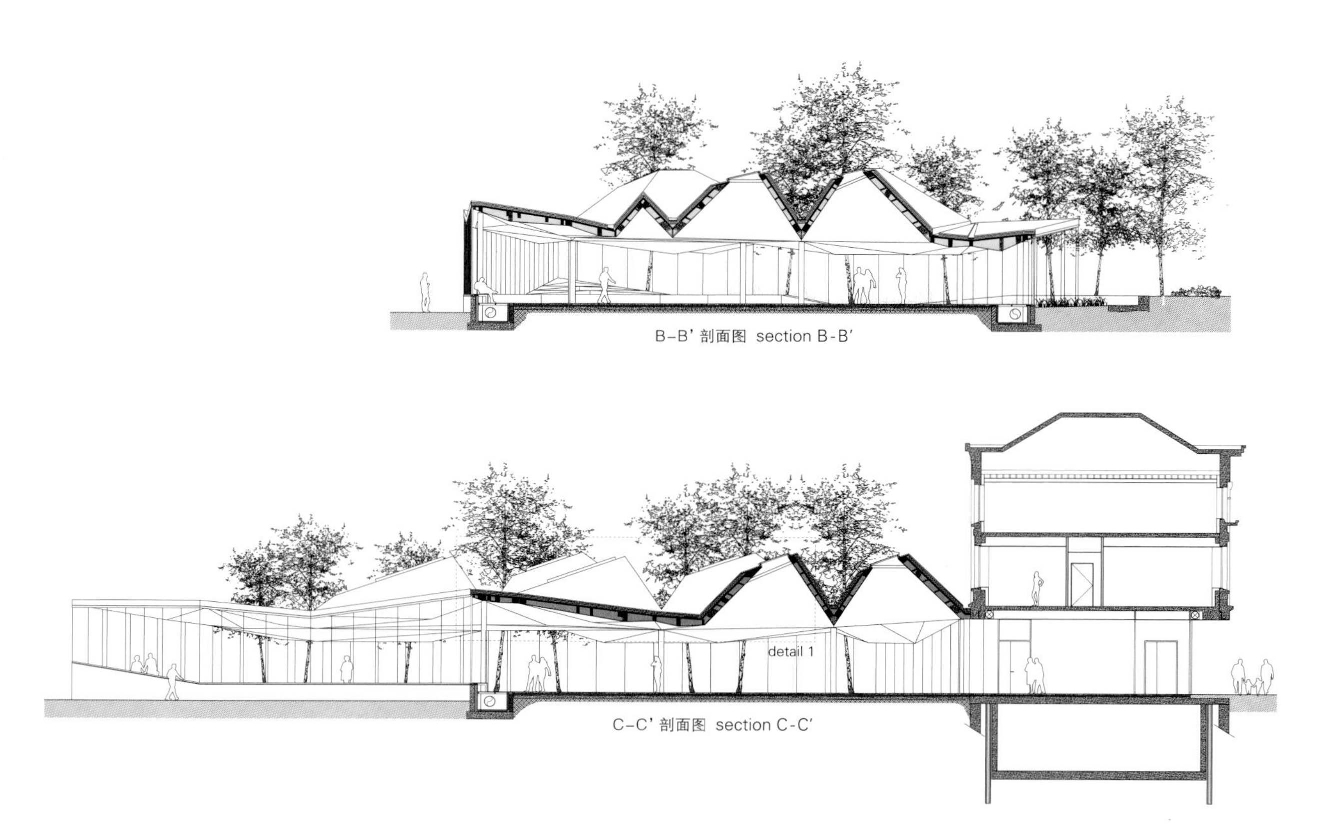

B–B' 剖面图 section B-B'

C–C' 剖面图 section C-C'

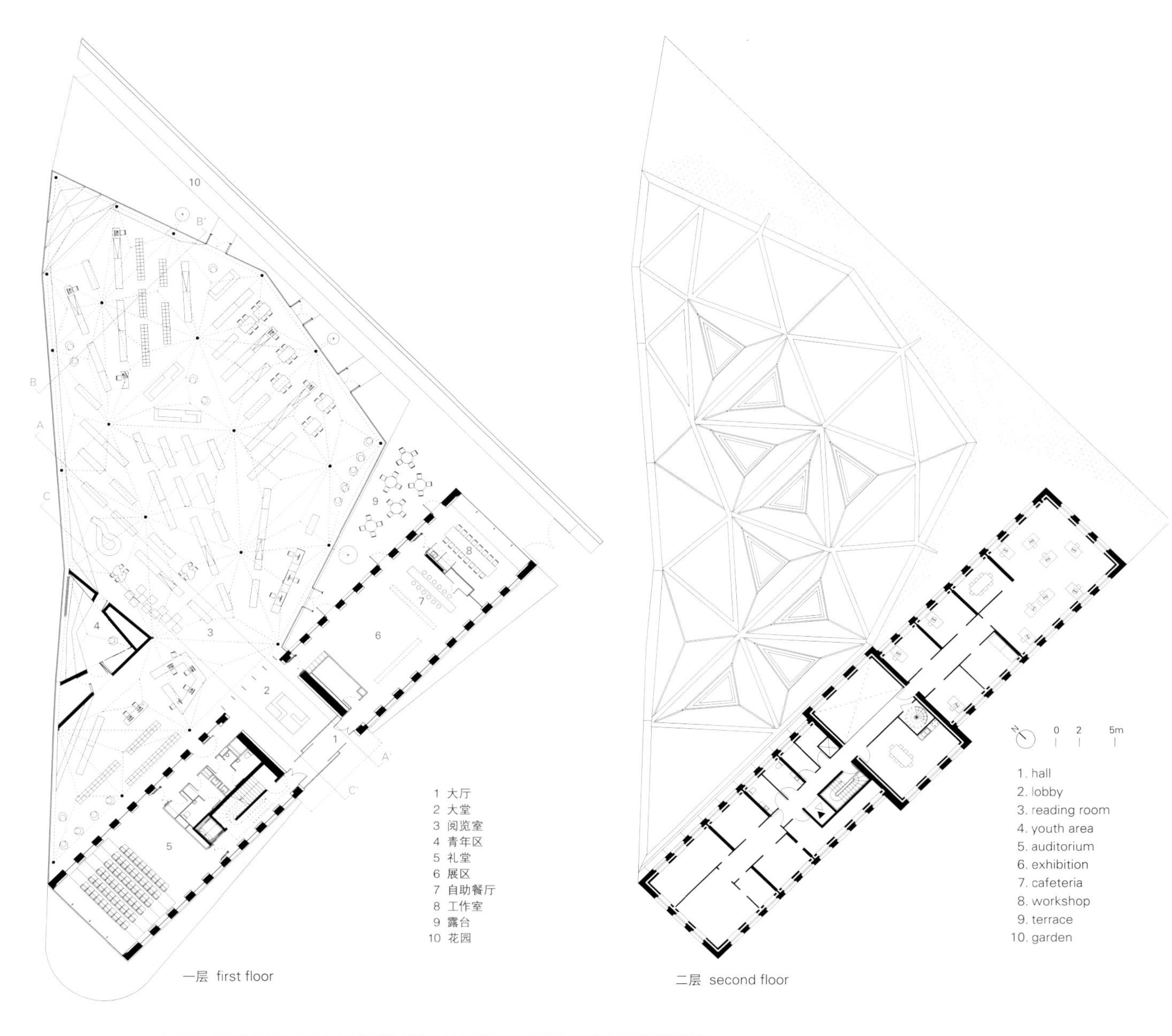

项目名称：Madeleine Media Library
地点：La Madeleine, France
建筑师：Tank Architectes
项目经理：Clément Berton
工程研究：SNC Lavalin
经济师：PHD Ingéniérie
音效师：Daniel Caucheteux
甲方：Ville de La Madeleine
用地面积：2,196m²
总建筑面积：1,800m²
有效楼层面积：2,200m²
竣工时间：2013.11
摄影师：
©Julien Lanoo (courtesy of the architect) - p.86, p.87, p.88~89, p.90, p.91, p.92, p.93, p.96, p.97, p.98, p.99
©Pierre Manuel Rouxel (courtesy of the architect) - p.94, p.95

level at the same level of the public space. The ground floor is designed as a continual space.
The existing building facing the market place organizes the entry sequence and deals with the public space. Here is an hall, an auditorium, an exhibition space, a cafeteria and a pedagogic workshop. This reception space naturally leads people to another universe much quieter: the extension. The timber ceiling punctuated by skylights seems to float above the ground. This space is the main reading that opens to the sky and adjacent streets and garden. The room is organized by roof variations which extend and compress the space that gives a more personal space. The result is a calm and peaceful atmosphere inviting to everybody access to a variety of media.
The building covers almost the entire land. Nine triangulated sheds are placed on a regular frame of columns, covered with glass oriented to the North to catch soft and diffused light. The roof is adapted to its environment opens onto the urban landscape and garden. It folds and softly goes down to guide water and pour it into the garden through three gargoyles. The 90 timber facets are used to create a variety of spaces in a fully open and flexible place.

Andrée Chedid媒体图书馆

D'HOUNDT+BAJART architectes & associés

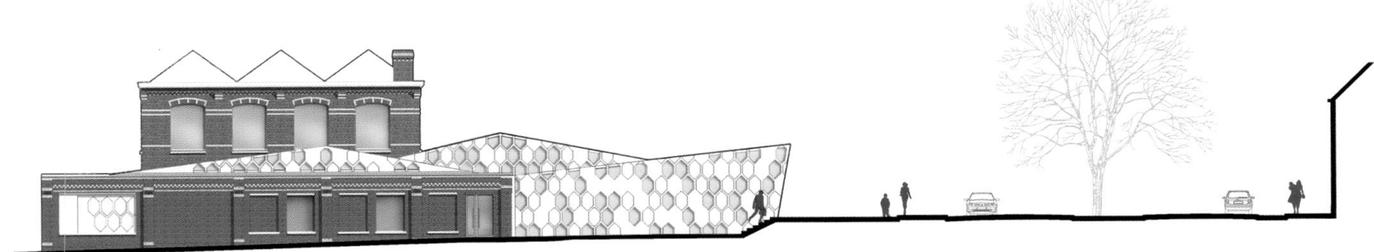

西立面 west elevation

该建筑的表面面积超过900m²，由两部分组成。原有部分是Tiberghien纺织厂的门房服务处及其围墙，扩建部分则与历史建筑元素唇齿相依，连为一体。新建筑由三个建筑体量组成，其西侧紧靠翻新的纺织厂围墙，东北部则与老旧的门房服务处连为一体。这三个体量的屋顶连为一个连续的表面，呈波纹状起伏，为嵌在媒体图书馆中央的不同区域提供所需要的自然光线。东面墙体和西面墙体为一系列穿孔结构，里面的人可以看到外面，但路人无法看到里面。建筑入口经过精心设计处理，直观而又明显，穿过门房服务处。该入口保存完好且进行了精雕细琢，经过切割后大面积向庭院开放。一层为公众服务区，二层专门留作媒体图书馆内部服务区。

像一个矿物雕塑

老Tiberghien纺织厂工业化的过去让人们记忆犹新，而新扩建的三栋建筑物犹如这座媒体图书馆长出的三个白色翅膀，其构造就像有机晶体矿物一样，与地形相得益彰，伸展开来，向上攀升，进入路人的视野，好像既要抓住人们的眼球，又要抓住光线。建筑立面非常醒目，毅

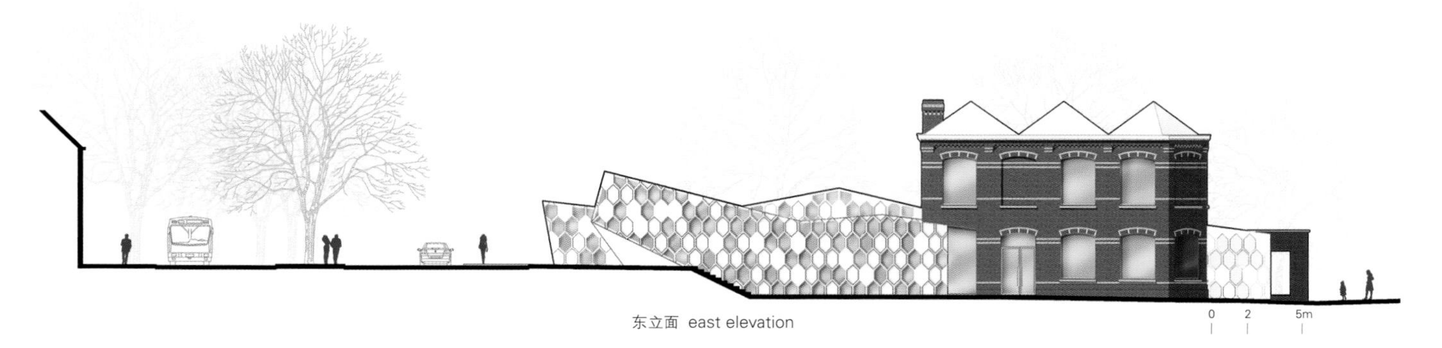

东立面 east elevation

然决然地展望着未来，而北面和南面是封闭的，东面和西面为开孔墙面，呈现了一种"内外翻转"的形式，暗示内外生活之间无形的联系。

内饰：文化与灯光的汇聚地

在这处开放、无拘无束的空间里，为了使空间特征与建筑语言保持高度一致，家具全部定制。白色或者木质的元素是为成年人准备的，而彩色的元素是为小孩准备的。进入室内的光线，特别是经过蜂房式开孔进入室内的光线，都能精确恰当地照射在设计区域：儿童角和阅览角。在媒体图书馆的核心区域，光线通过波浪状起伏的屋顶照射进来。地面铺的是拼块地板：地板由一块块外形不一的板块拼贴组成，可以灵活组成不同样式和图案，打破了以前单一死板的铺设方式。不同的地块拼接到一起，不会出现重复的图案，不同区域可拼接不同地毯，在视觉上与媒体图书馆的有机晶体矿物结构相呼应。

发挥木质材料无限潜能

用于外立面和屋顶的木质结构采用传统的建筑工艺，体现了高度

精确的3D数字设计拼接技术。该建筑就用这种方法来定义其外观轮廓，堪称史无前例，无与伦比。

媒体图书馆的框架是真正在工厂车间里预安装的原型，建成蜂巢形状的外立面和屋顶最后使用了3120块木材单元，以及2134块金属组装单元。

Andrée Chedid Media Library

The building will be developed over a surface area of 900m² and will consist of two parts, an old part – the Tiberghien textile factory's concierge service along with its surrounding wall – and an extension that will lean on these historic elements.

The new building, leaning against the surrounding renovated wall to the west, and against the old concierge service to the north-east, will consist of three volumes, where the roofs, treated as a continuous surface, will ripple to bring the natural light needed for the different spaces nestled in the heart of the media library. The eastern and western walls will form a set of perforated structures, opening the view to the outside while protecting it from the passer-bys looking in. Attention is paid to the entrance treatment in an intuitive and evident way, through the concierge service, preserved yet carved, sliced and widely opened onto the courtyard. The ground floor will host the public, since the first floor is reserved for the media library internal services.

Like a Mineral Sculpture

From an industrial past that is still very present in the collective memory – the old Tiberghien spinnings arise three white wings from the media library. Their structures, similar to an organic matter, match the ground shape, spread out, climb and impose to the passers-by's view, as if to catch both their eye and the light. This building, resolutely looking to the future, solicited by its remarkable elevations, closed from the north

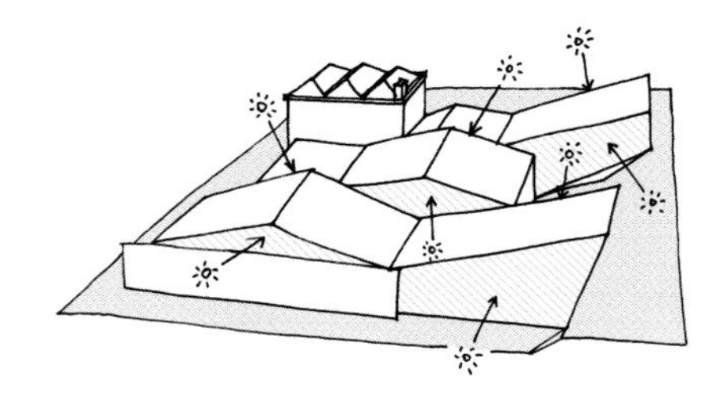

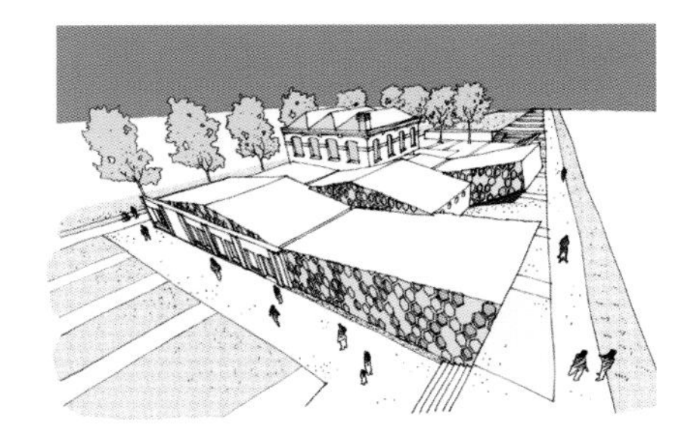

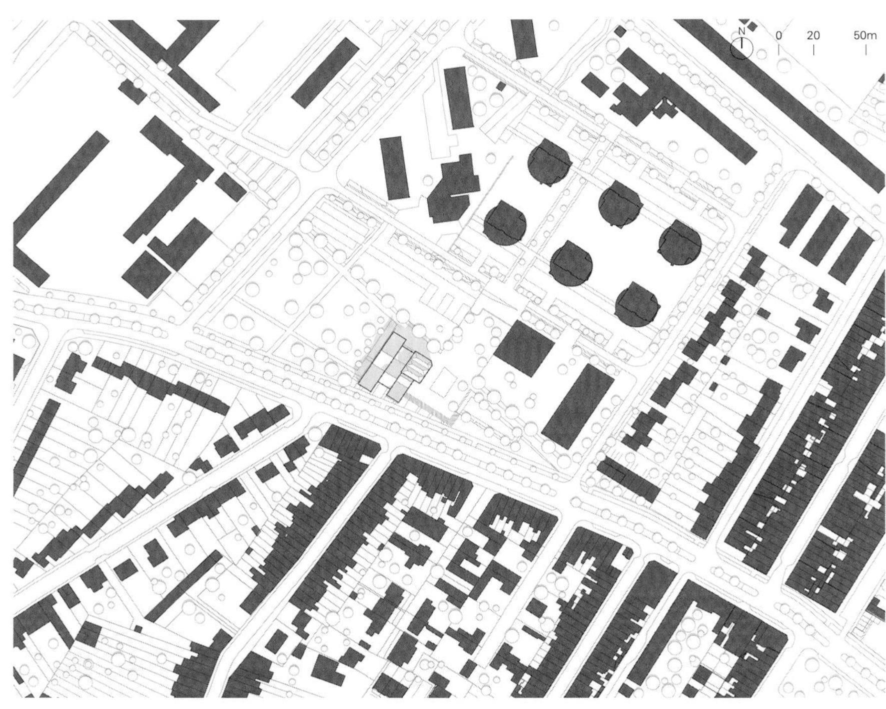

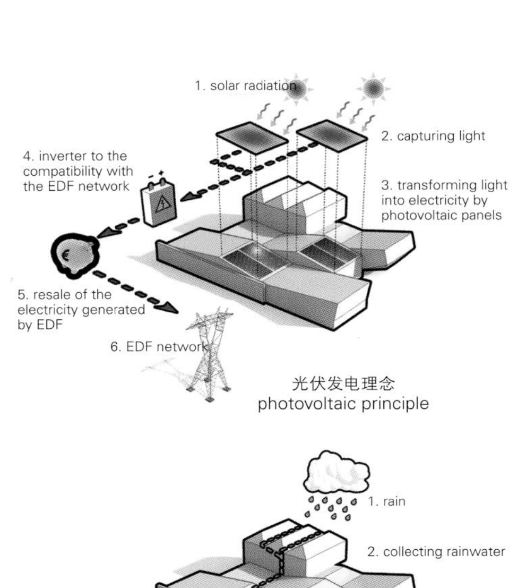

光伏发电理念
photovoltaic principle

雨水回收理念
rainwater recovery principle

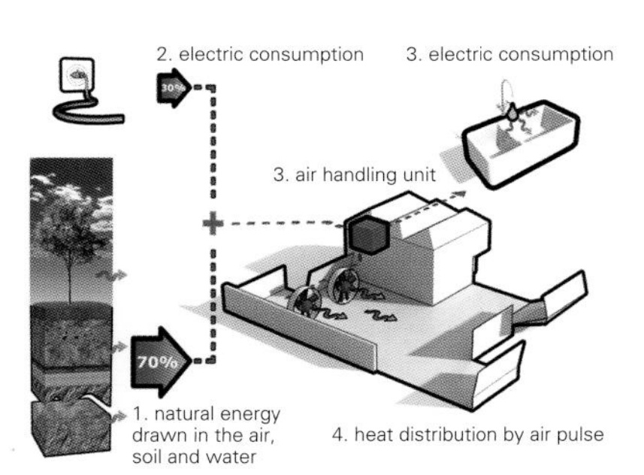

产品加热理念
production heating principle

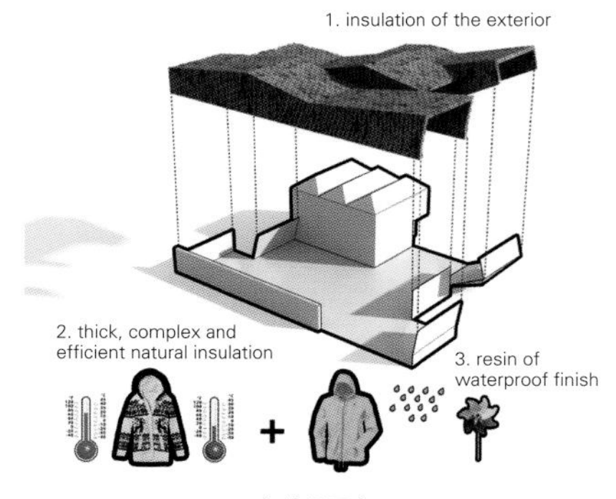

与外界隔离
isolation from outside

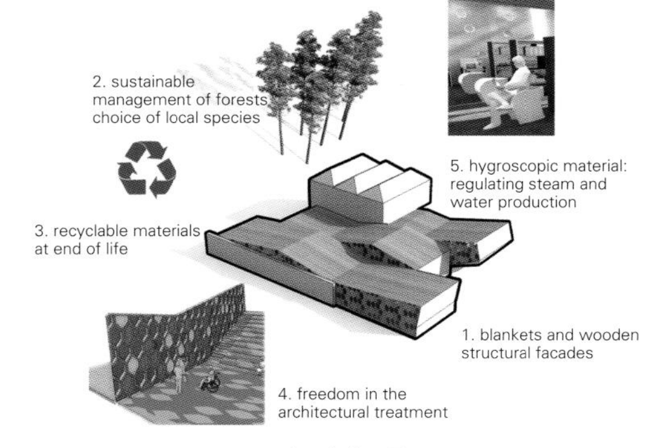

立面和木覆层
facades and wooden cover

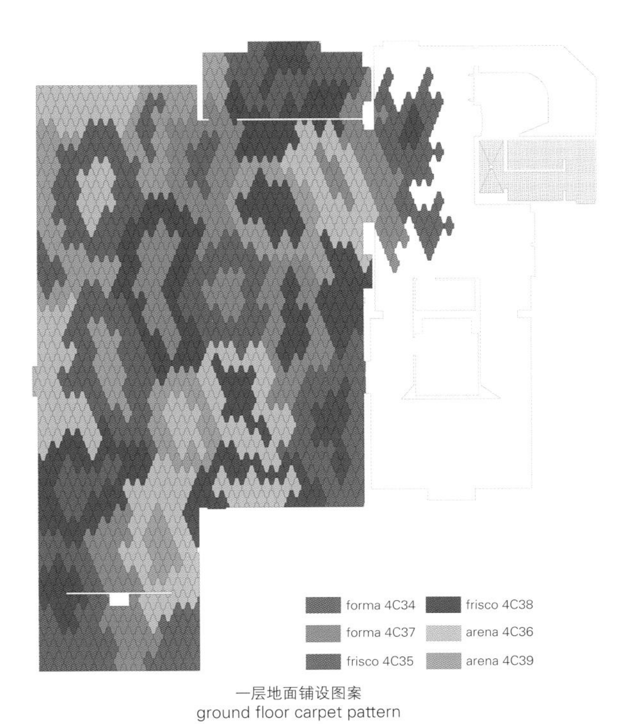

一层地面铺设图案
ground floor carpet pattern

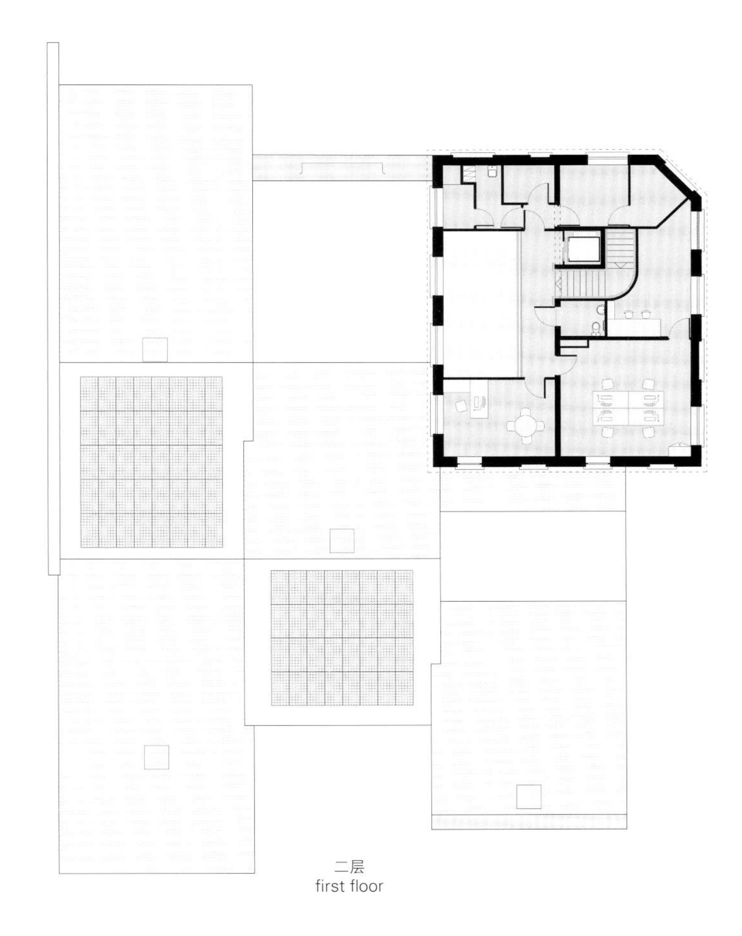

二层
first floor

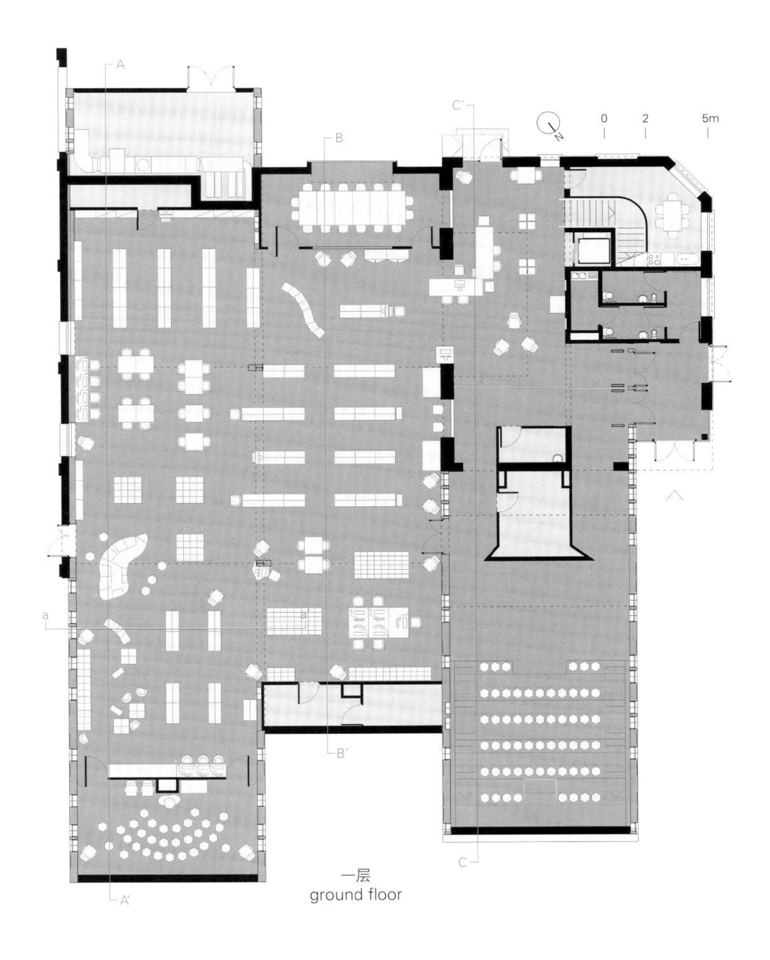

一层
ground floor

项目名称：Andrée Chedid Media Library
地点：Tourcoing, France
建筑师：D'HOUNDT+BAJART architectes & associés
业主：Ville de Tourcoing
立面设计：VS-A group
功能：media library, auditorium
有效楼层面积：920m²
造价：EUR 2,376,500
竣工时间：2013.9
摄影师：courtesy of the architect

and the south, opened in mesh from the east and the west, provides an "inside out" form suggesting the impalpable link between the inner and outer life.

Interior: A Place of Cultures and Lights

In this open and liberated space, the furniture has been custom designed to fully participate in the identification of areas for the sake of the architectural consistency. White or Wood is used for adults, while color is used for the younger. The light participates in the interior especially with the cells that allow a precise and directed input on selected areas: the children's corner and the reading corner space. At the heart of the media library space, it is through the roof undulations that the light penetrates. The flooring consists of carpet tiles: a carpet of free-form tiles which allows great flexibility in patterns creation, and breaks with the rigid structure usually proposed. The interlinked parts generate non repetitive graphics, specific to different areas, and form a visual echo to the organic structure of the media library.

Pushing the Limits of the Wood Material

The woodwork, for the facade and the roof, signs the junction of the 3D digital design, highly accurate, with the traditional building technique. This building, one of a kind, defines the outlines of a high precision wood architecture never achieved to date.

The media library skeleton, a real prototype pre-assembled in the workshop, needed 3,120 pieces of wood in order to complete the honeycomb facades and the roofs, and 2134 assembly metallic pieces.

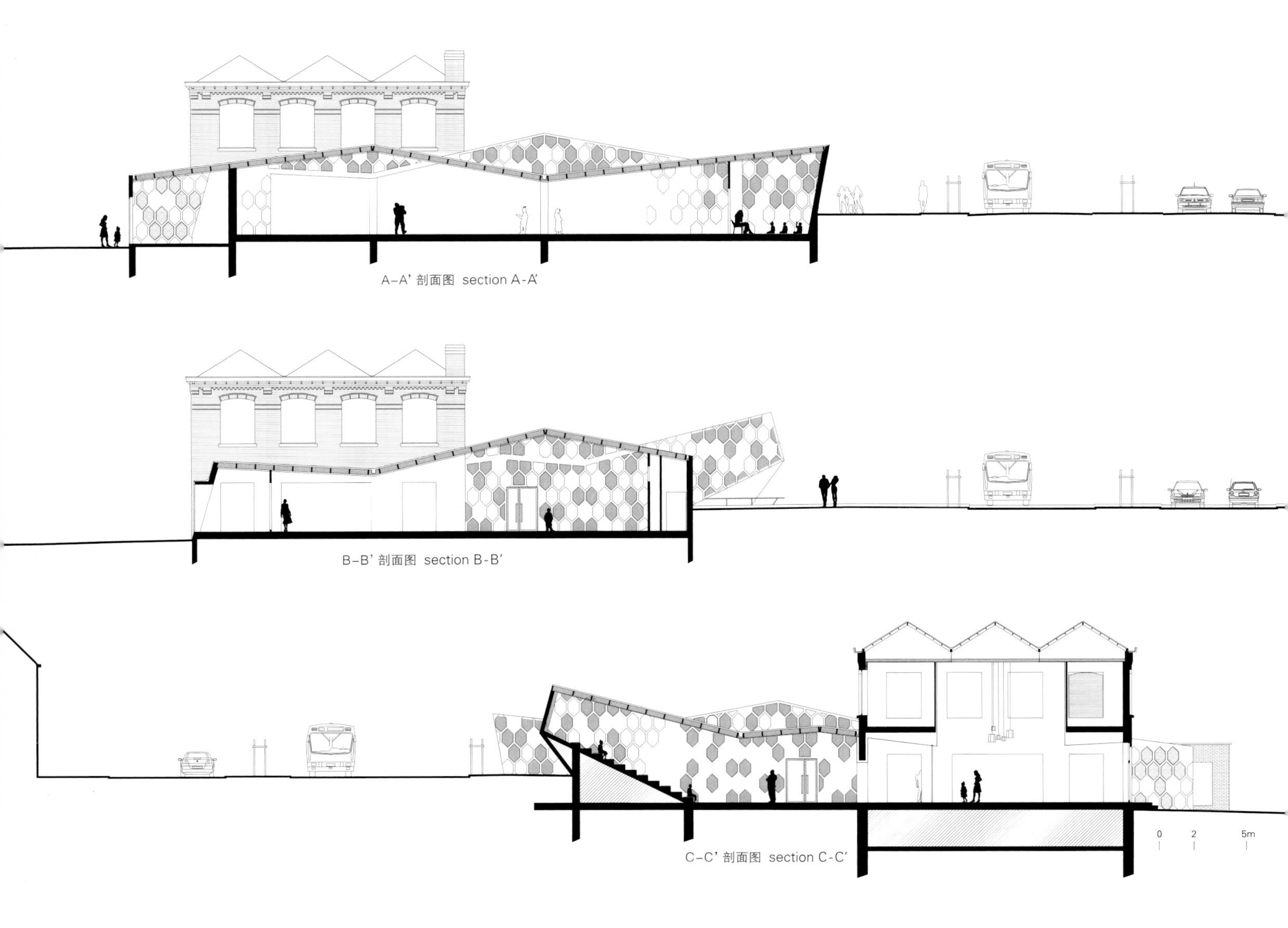

A-A' 剖面图 section A-A'

B-B' 剖面图 section B-B'

C-C' 剖面图 section C-C'

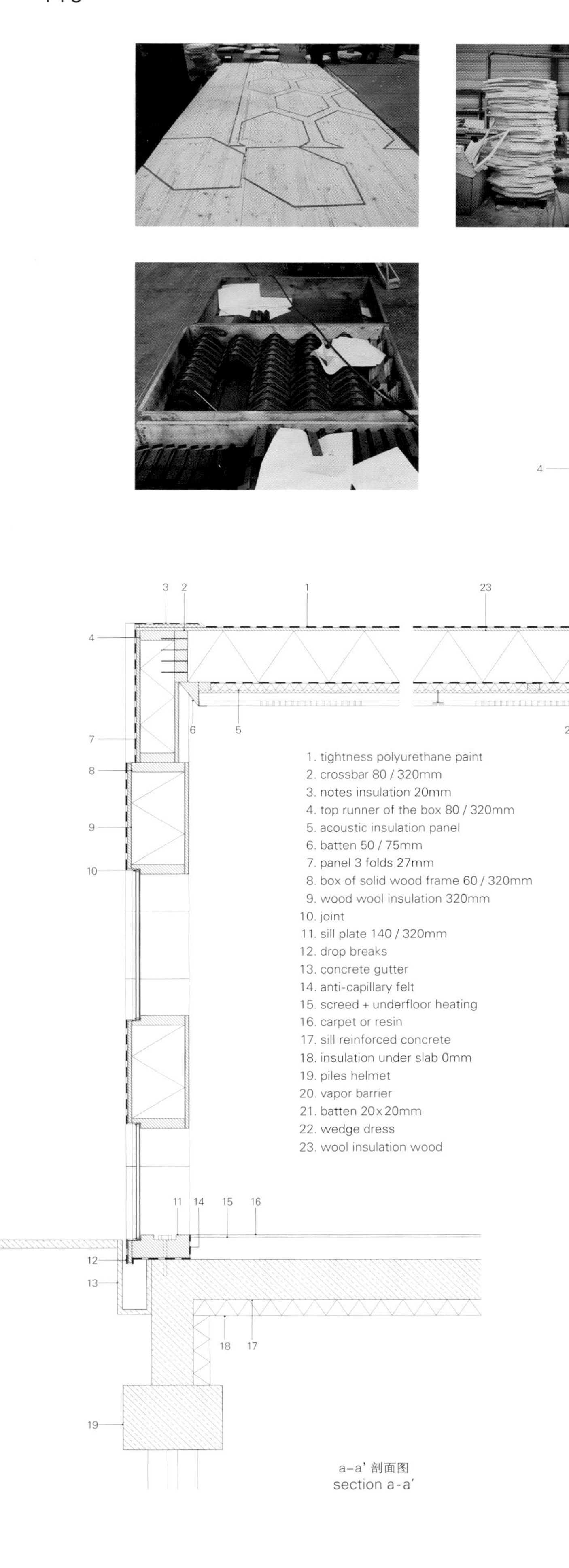

a-a' 剖面图
section a-a'

微院儿

ZAO/standardarchitecture

茶儿胡同（茶胡同）是嘈杂忙碌的大栅栏地区中一个安静的所在，距离市中心的天安门广场有1km。茶儿胡同8号是典型的"大杂院"（大–杂乱的–庭院），曾有十多户人家在此居住。在过去的五十年里，每一家都在院子里搭建了一个小厨房。这些违章搭建的建筑让这个地方特别密集。在过去几年的城市整修改造中，这些违章搭建的建筑几乎都被拆除了。

在这个项目中，建筑师试图重新设计、改造和重新利用这些非正式的附加建筑，而不是拆除他们。这样做，他们意在把这些附加结构看做一个重要的历史见证，是北京胡同中经常被忽视的现代平民生活的重要体现。

这个主要由胶合板建造的儿童图书馆只有9m²，镶嵌在原有建筑物的坡屋顶下面，与仍然住在院子里的人家共生共栖。大白蜡树下以前曾经是一户人家的厨房，现在被重新设计成了一个6m²的迷你美术空间，由传统的蓝灰色砖块建造。在这个结构外部有一个砖质楼梯，可以通向屋顶。在那里，白蜡树的树叶和枝干伸手可及。通过在茶儿胡同的庭院中创造小型建筑结构，建筑师们试图加强社区之间的联系，丰富当地居民的胡同生活。孩子们放学后，在他们的父母来接他们之前，可以在这里挑一本自己喜欢的书，找一个属于自己的小角落浏览一会儿。或者，孩子们可以爬到楼顶，坐在树荫下，在熟悉但又独出心裁的环境中与长者温馨而亲切地交流。

Micro-Yuan'er

Cha'er Hutong (Hutong of tea) is a quiet spot among the busy Dashilar area, situated one kilometre from Tiananmen Square in the city centre. No.8 Cha'er Hutong is a typical "Da-Za-Yuan" (big-messy-courtyard) once occupied by over a dozen families. Over the past five decades each family built a small add-on kitchen in the courtyard. These add-on structures form a special density that is usually considered as urban scrap and almost all of them have been automatically wiped out with the renovation practices of the past years.

In this project the architects tried to redesign, renovate and re-use the informal add-on structures instead of eliminating

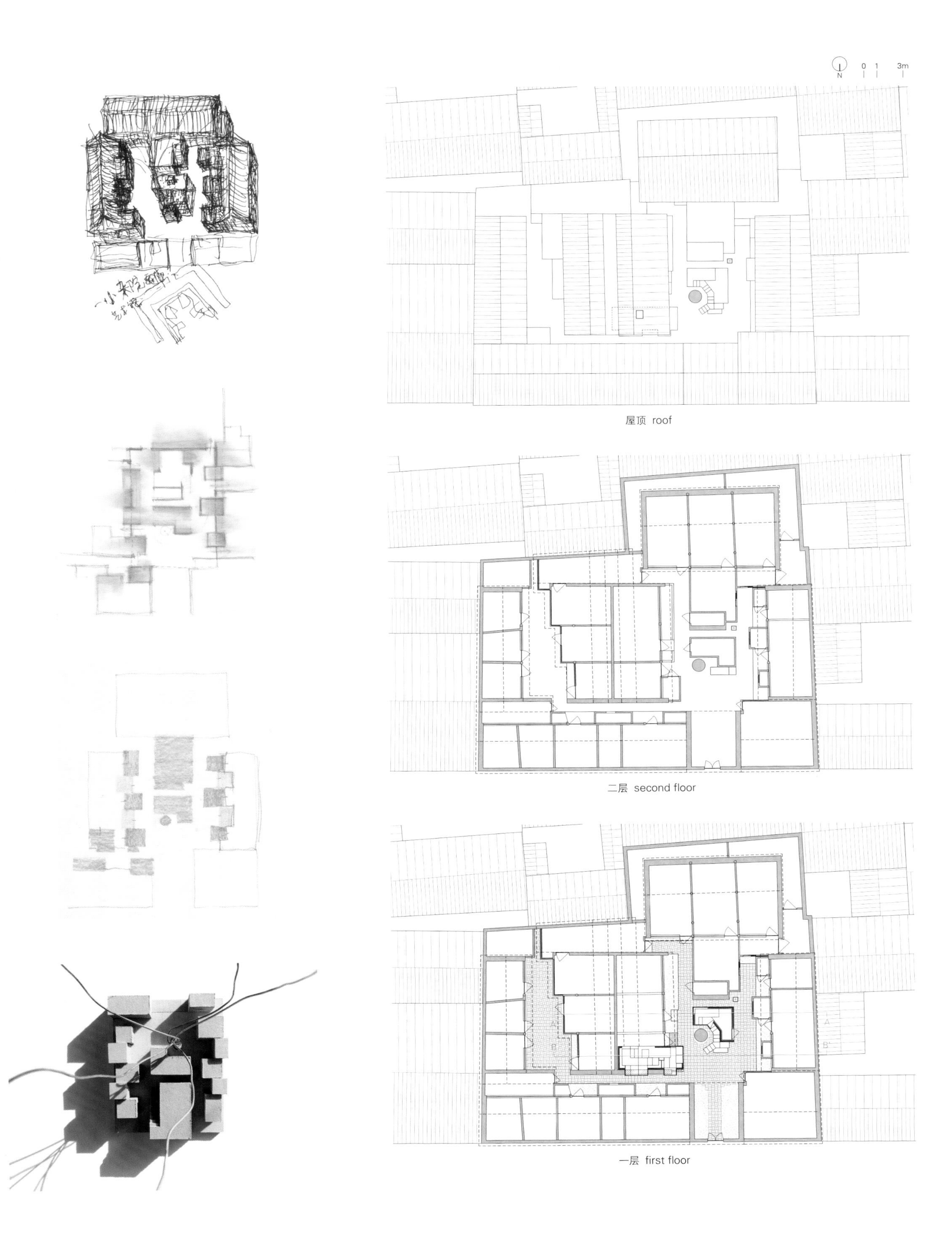

屋顶 roof

二层 second floor

一层 first floor

项目名称：Micro-Yuan'er
地点：Cha'er Hutong #8, Dashilar, Beijing, China
建筑师：Zhang Ke, Zhang Mingming, Fang Shujun _ ZAO/standardarchitecture
设计团队：Ao Ikegami, Huang Tanyu, Dai Haifei, Zhao Sheng, Liu Xinghua
赞助商：Camerich
有效楼层面积：8m²_art space, 9m²_library
设计时间：2012.6~9 施工时间：2014.9 (12 days) / 竣工时间：2014
摄影师：
©Su Shengliang (courtesy of the architect)-p.112~113, p.115, p.116, p.117, p.118, p.119, p. 120, p.123bottom
Courtesy of the architect)-p.121, p.123top

them. In doing so, they intend to recognize the add-on structures as an important historical layer and as a critical embodiment of Beijing's contemporary civil life in hutongs that has so often been overlooked.

In symbiosis with the families who still live in the courtyard, a nine-square-metre children's library built of plywood was inserted underneath the pitched roof of an existing building. Under a big ash tree, one of the former kitchens was redesigned into a six-square-metre mini art space made from traditional bluish grey brick. On its exterior, a trail of brick stairs leads up to the roof, where one may delve into the branches and foliage of the ash tree. With the small-scale intervention in the Cha'er Hutong courtyard, the architects try to strengthen bonds between communities, as well as to enrich the hutong life of local residents. A child may stop by after school, pick out a favourite book, and read in his little niche before getting picked up by the parents. Or the kids may climb up onto the roof, sit in the shade, and engage in a cosy conversation with the elderly in a familiar but new space.

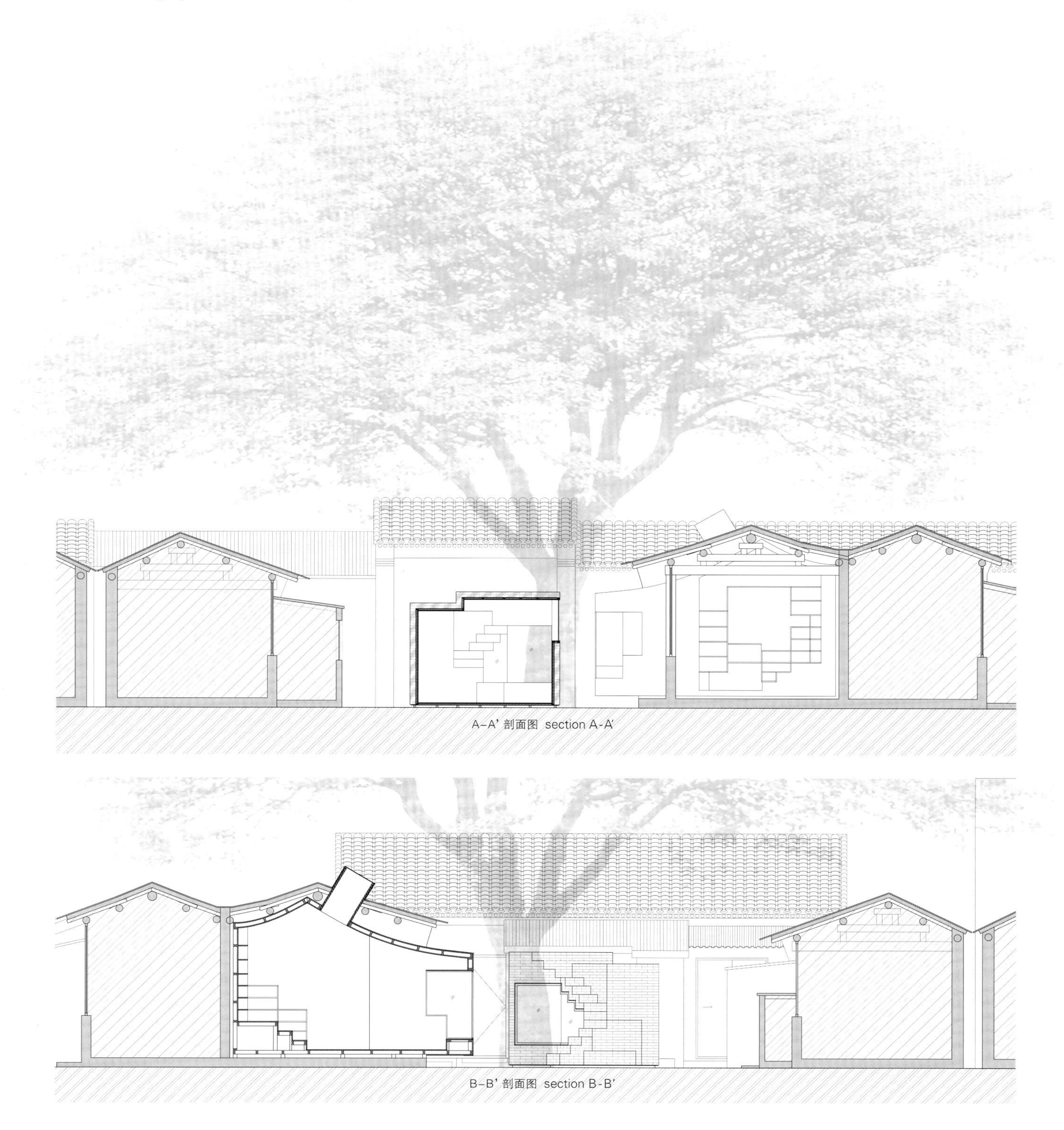

A-A' 剖面图 section A-A'

B-B' 剖面图 section B-B'

方格子老虎
会说晚安的故

机器是
怎样
工作的？
透了!

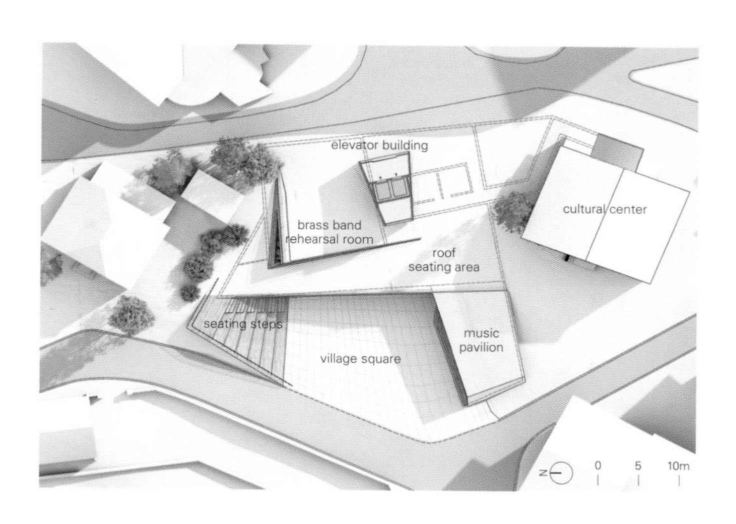

伊施格尔镇圣尼古拉斯文化中心

Parc Architekten

伊施格尔镇是阿尔卑斯山最重要的旅游目的地之一。小镇居民有1600人，可接纳游客11 000人。旅游旺季几乎可持续半年的时间。在旺季，小镇到处都是滑雪者，呈现一片狂欢的景象。滑雪后的娱乐社交活动似乎比在滑坡上的体育活动还要重要。

这座小镇在过去的30年中迅速发展成为宾馆聚集区。虽然如此，小镇作为一个社交场所的意义仍然存在，镇里的人们组织了40多个非常活跃的俱乐部。这样做或许是为了响应如火如荼的旅游业。比如说，铜管乐队就有100名活跃成员（小镇只有1600位居民），在这一地区相对来说是最大规模的。当地居民之间的联系越来越密切，有些地方对他们来说非常重要。这些地方既不是为了游客而建，也不是为了赚钱而建。

2012年，伊施格尔镇决定只为当地居民建造一座文化中心。教堂和"新"与"旧"的牧师之家之间还有一片草地，这在以前的乡村里相当普遍。镇政府希望该建筑成为伊施格尔镇"身份"的象征，一方面可以体现伊施格尔镇自信、成功的今天，另一方面可以让人们感受到伊施格尔镇绿意盎然的往昔。怀着这样的愿望，伊施格尔镇举行了建筑设计竞赛。

这座新建筑主要位于地下，铜管乐队在此排练演出。合唱团工作室、图书馆和小镇档案馆便位于其中的一个老牧师之家。有简单的烹饪设施的休息厅可供所有的俱乐部使用，但没有像在"正式的"酒吧那里必须消费的压力。场地的下面横穿着原有的隧道，可以把宾馆的游客送到去滑雪的电梯里。

新建筑所在区域就是隧道的单行通道口，所以整栋建筑就围绕通道口的电梯而建，电梯所在的建筑就像这一新建景观内的一块巨石。建筑的某些部分（主要由混凝土建造）突出地表，其玻璃表面允许光线照射进来。草地层的下方有一个小村庄广场。在举行活动的时候，玻璃立面可以打开，这样门厅可以用作不受天气影响的餐厅。草地的设计自然而流畅，将村庄广场和建筑物的顶部连为一体。音乐排练室是这座建筑最核心的部分，就像嵌进半山腰的木盒子。墙与墙之间都有一个轻微的角度，这样可以防止颤动回声。五百个三角形折叠式吸音木板能够使这个房间达到声音效果方面的要求。窗口区域前面的嵌板可以打开，让光线照进来。

乐队演奏台位于镇广场的最南端，后墙由符合声学原理的木构件制成。周围的玻璃护栏可以阻挡来自山上的冷风。乐队演奏台成为村庄广场的延伸，可供人们永久使用。与这一乐队演奏台相对的空间是对面带有木质看台座位的斜坡。

人们可通过地下通道和新建的楼梯井/电梯间进入列管的"牧师之家"。地下部分都是钢筋混凝土结构，而地上部分是传统的实木结构。牧师之家的外观非常朴素简单，但是里面几乎展示了18世纪一个富有牧师的家的原貌。翻新重建工作大规模地展开，再现的颜色层、清洁旧的木板、门以及带有横梁的天花板，使原建筑的巴洛克式风格一一展现出来。

St. Nikolaus Cultural Center in Ischgl

Ischgl is one of the most important tourism destinations in the Alps, with 1,600 residents facing over 11,000 guest beds. In the season, that lasts almost half a year, the townscape is dominated by skiers in best party mood; the apres-ski party seems to be more important than the sport activities on the slopes. The small town has grown in the last 30 years to a huge cluster of hotels. Nevertheless, the town still exists in a social sense, and people are organized in more than 40 clubs which are very active, perhaps also in response to the overwhelming tourism. For example, the brass band with almost 100 active members is (for a village with 1,600 inhabitants) relatively the largest in the region; the native people move closer together and there are places important that are not meant for tourists and for earning money.

In 2012, Ischgl has decided to build such a building only for the residents. Between church and "old" and "new" Widum(the house of the priest) there was still a rest of the meadow which was once typical for this type of rural villages. The municipality wanted an "identity creating building" that corresponds on the one hand with the self-confident and successful Ischgl of today, yet still leaves the ability to sense the green village of bygone days. With that wish they held an architectural competition.

The new building is mainly built underground. It will be used

Valülla

by the brass band, while choir room, library and archivist of the town find place in one of the old "Widum". A lounge with simple gastronomic facilities is usable by all clubs, but without the pressure of consumption like in a "normal" bar.

Under the site crosses an existing tunnel that transports guests from the hotel to the ski lifts.

In the area of the building is one access-point to the tunnel. So the house was built around this lift, and the lift access-building sits like a boulder on the new created landscape. Some parts of the building (mainly made of concrete) break through the surface and glass surfaces allow light to penetrate. At the foot of the meadow, a small village square is located. For events, the glass facade can be opened and the foyer then serves as a weather-protected culinary supply. A naturally designed way on the meadow connects the village square with the top place on the building.

The music rehearsal room is the centerpiece of the building, like a wooden box pushed into the hillside. The walls are set at a slight angle to each other, which avoids flutter echoes. 500 triangular foldable acoustic panels of wood allow an acoustic matching of the room. The panels in front of the window area are opened and allow light to penetrate.

The bandstand is the southern termination of the village square. The rear wall is formed by acoustic wooden elements, and a glazing on the terrain line protects from the cold wind from the mountain. It forms a permanently usable expansion of the village square. The spatial counterpart of the pavilion is the rising ramp with seats wood steps.

Access to the listed "Widum" is carried out via an underground connection and a newly built stairwell / elevator building. The subterranean elements are made in reinforced concrete, and the aboveground components in traditional solid wood construction. The Widum has a very unpretentious appearance, but inside it shows the nearly original conditions of the home of a wealthy priest of the 18th century. Extensive restoration was undertaken: the uncovering of old color layers, the cleaning of the old wood panels, doors and beamed ceilings provided the restoration of the Baroque inventory.

项目名称：Kulturzentrum Ischgl
地点：Eggerweg 4, 6561 Ischgl, Austria
建筑师：Parc Architekten
项目团队：Michael Fuchs, Barbara Poberschnigg, Di Dipl W-ing
合作者：DI Thomas Feuerstein, DI Thomas Feuerstein
用地面积：1,917m²
总建筑面积：1,200m²
有效楼层面积：2,500m²
设计时间：2012
竣工时间：2013
摄影师：©David Schreyer (courtesy of the architect) (except as noted)

HOTEL GARNI
Valülla

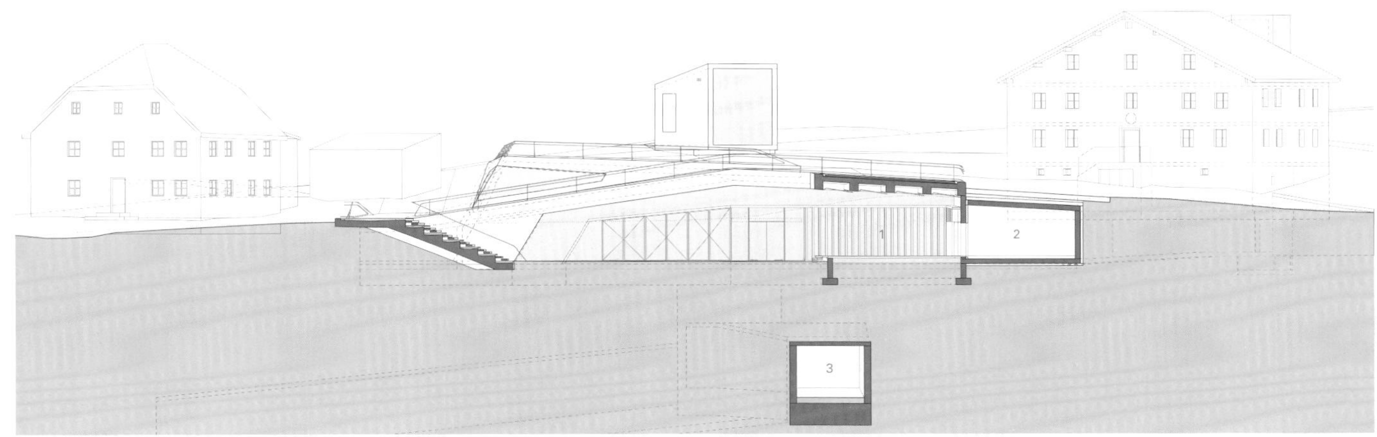

1 舞台 2 存储室 3 村庄的隧道 1. stage 2. storage 3. village tunnel
A–A' 剖面图 section A-A'

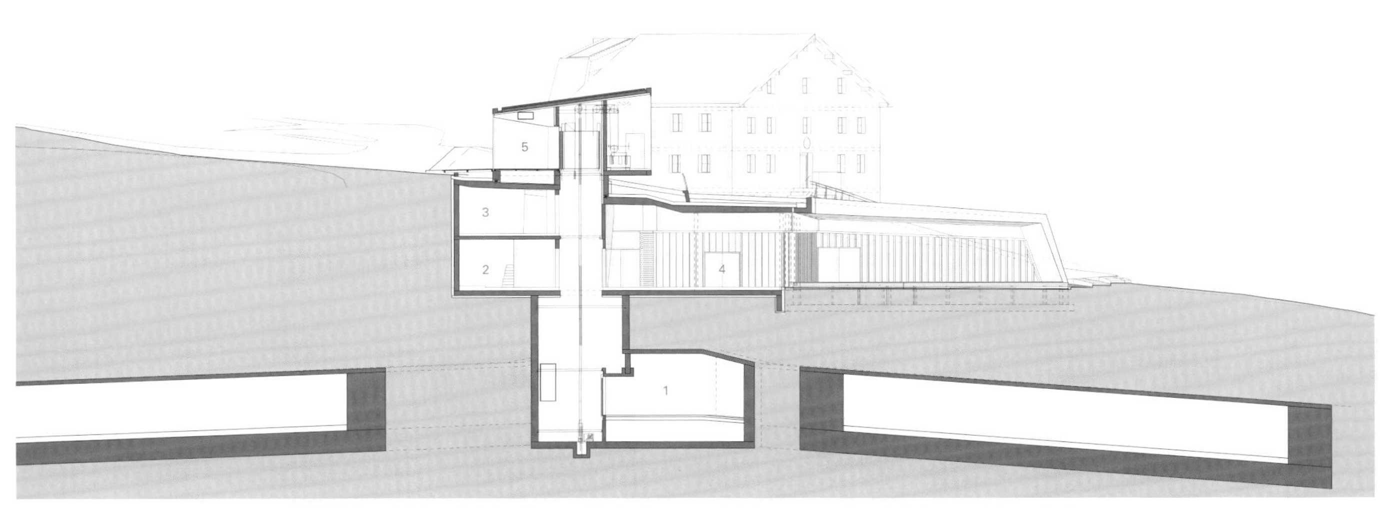

1 村庄的隧道 2 乐器室 3 技术室 4 门厅 5 电梯 1. village tunnel 2. instruments 3. technical room 4. foyer 5. elevator
B–B' 剖面图 section B-B'

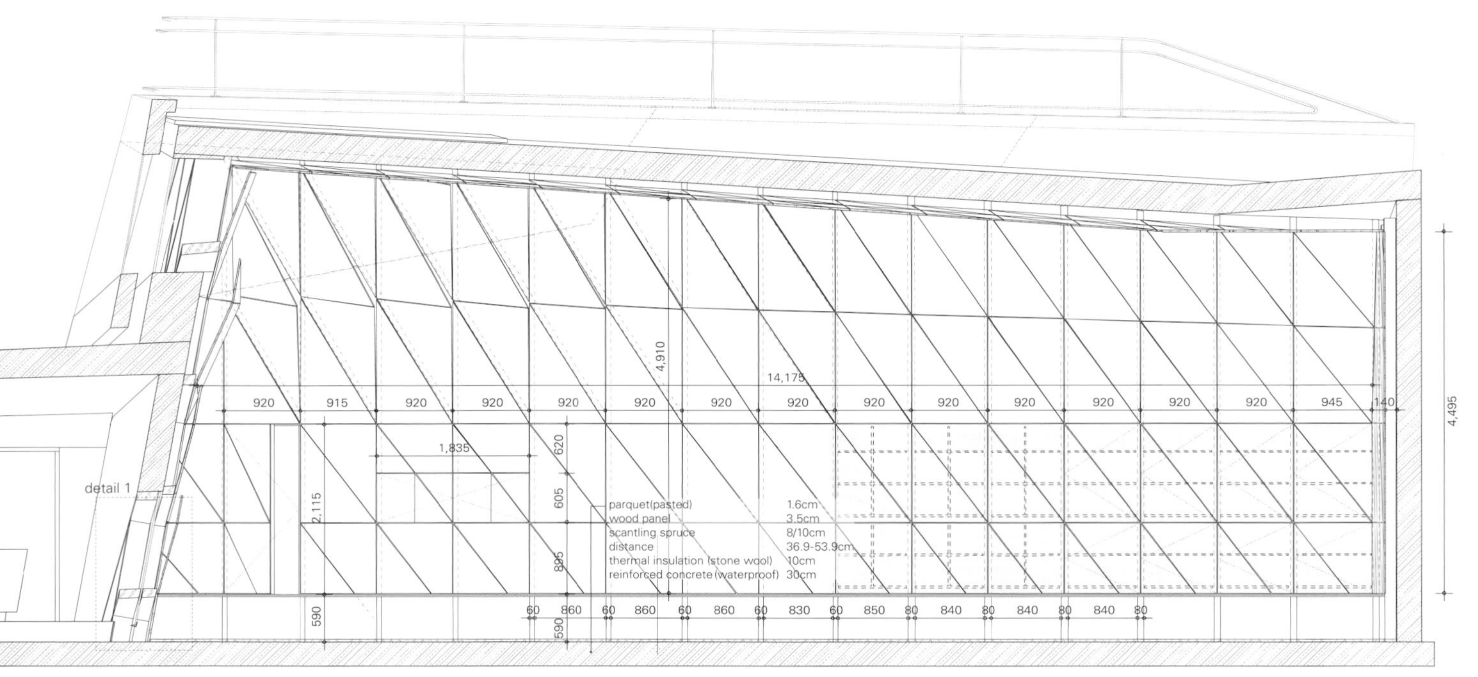

a–a' 剖面图 section a–a'

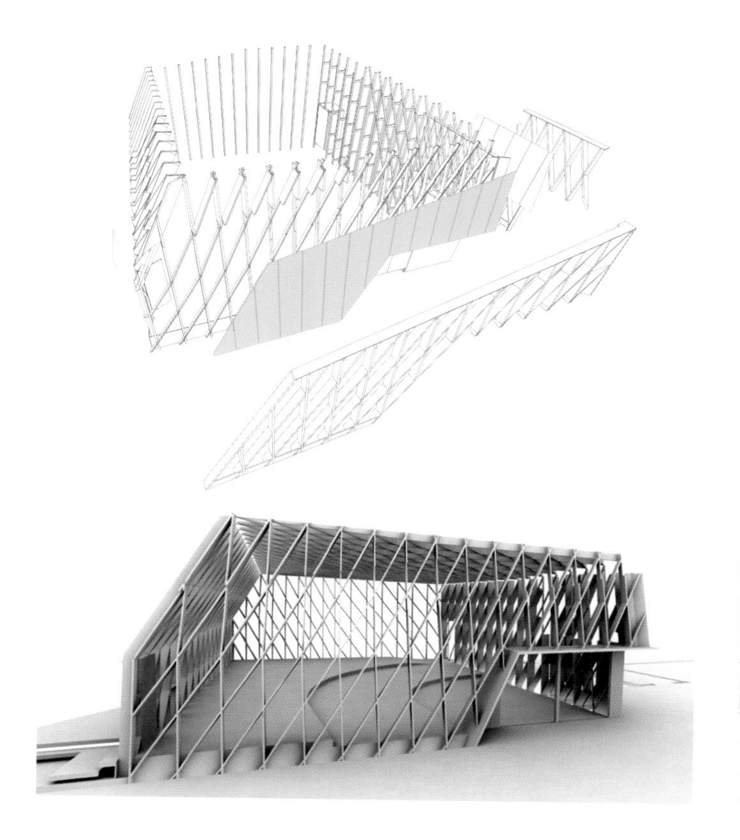

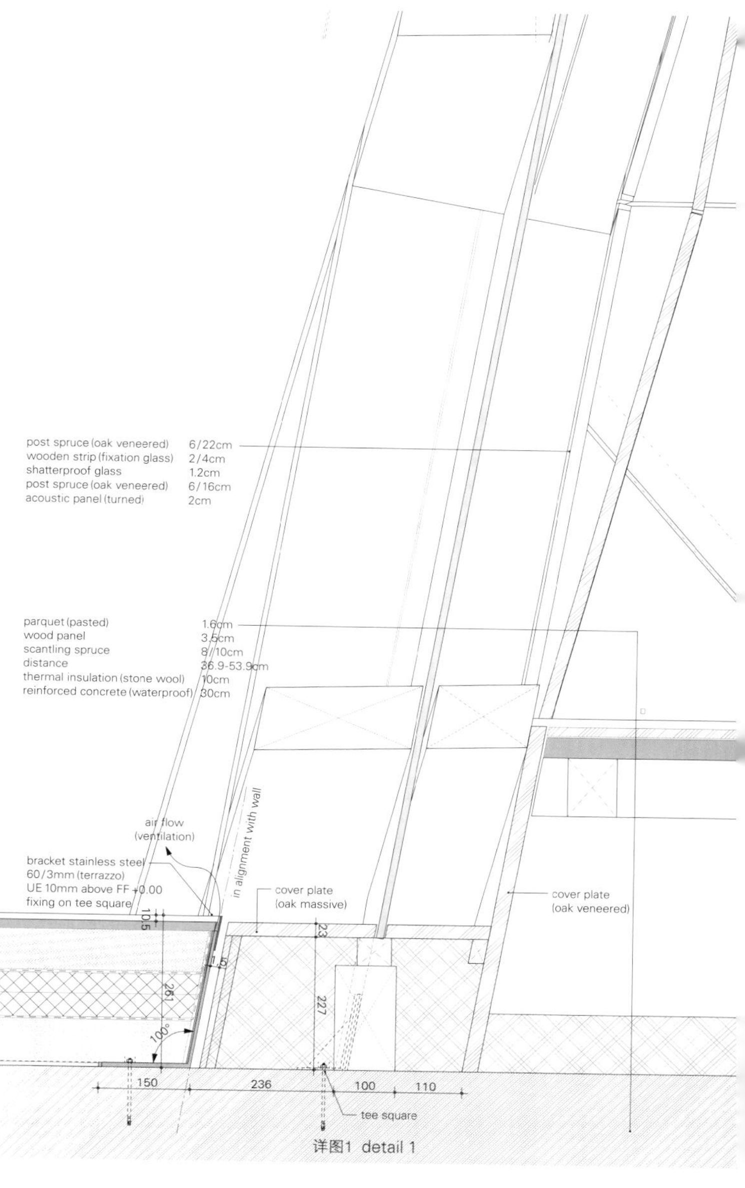
post spruce (oak veneered) 6/22cm
wooden strip (fixation glass) 2/4cm
shatterproof glass 1.2cm
post spruce (oak veneered) 6/16cm
acoustic panel (turned) 2cm
parquet (pasted) 1.6cm
wood panel 3.5cm
scantling spruce 8/10cm
distance 36.9-53.9cm
thermal insulation (stone wool) 10cm
reinforced concrete (waterproof) 30cm
air flow (ventilation)
bracket stainless steel 60/3mm (terrazzo)
UE 10mm above FF +0.00
fixing on tee square
in alignment with wall
cover plate (oak massive)
cover plate (oak veneered)
10.5
261
100°
23
227
150
236
100
110
tee square

详图1 detail 1

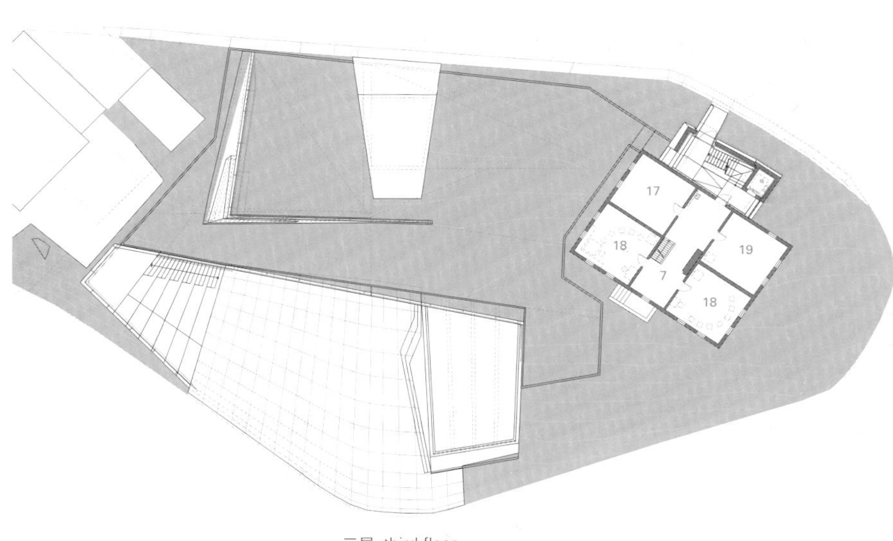

三层 third floor

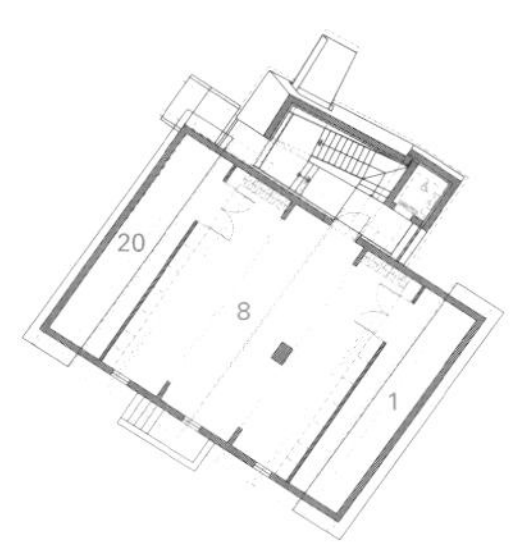

四层 fourth floor

1 存储室	1. storage
2 舞台	2. stage
3 社区	3. community
4 办公室	4. office
5 门厅	5. foyer
6 排练室	6. rehearsal room
7 通道	7. aisle
8 合唱团排练室	8. choir rehearsal
9 乐器室	9. instruments
10 杂物房	10. utility room
11 电梯	11. elevator
12 技术室	12. technical room
13 编年史作者工作室	13. annalist room
14 休息厅	14. lounge
15 阅览室	15. reading room
16 图书馆	16. library
17 传统服装存储室	17. storage for traditional costumes
18 音乐演练室	18. music practice room
19 乐器存储室	19. instrument storage
20 坐椅区	20. chairs

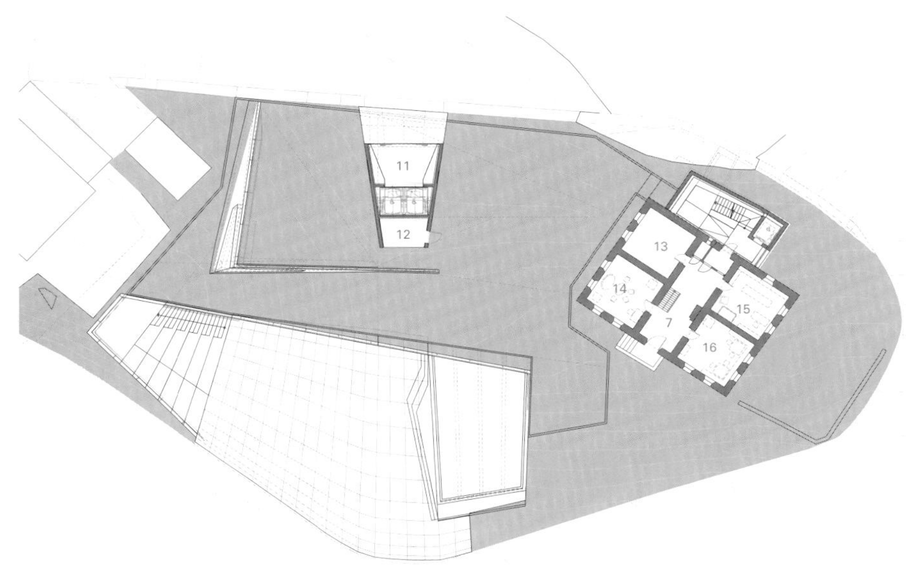

二层 second floor

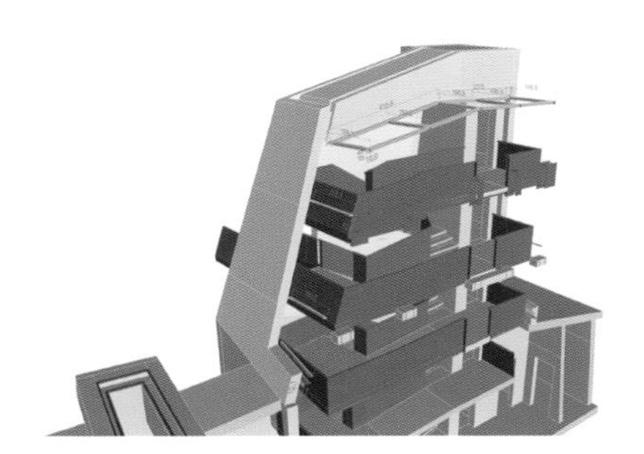

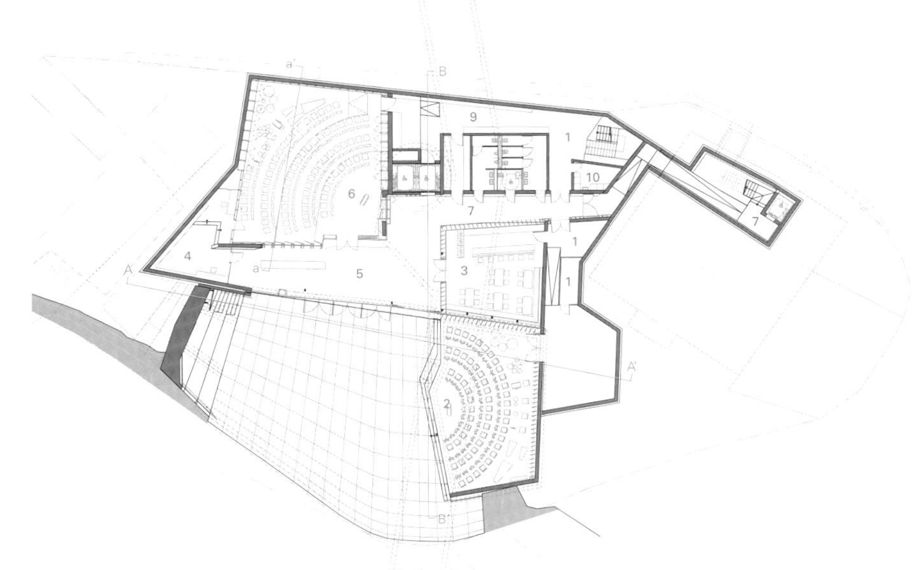

一层 first floor

照片提供©Karl Heinz (courtesy of the architect)

巴尼奥斯镇社区中心

Brown Meneses Arquitectos

为了满足重建厄瓜多尔文化遗址的需求，负责遗址的工信部命其应急管理部门找来了专业服务团队为巴尼奥斯最古老的教堂建造一个新屋顶。这些遗址包含了当地和区域发展的关键因素。

这座巴尼奥斯最古老的教堂毗邻现在的镇政府（建于1788年）。它于1797年遭地震破坏后修缮一次，但被1949年的地震破坏后彻底成为废墟。

人们从这一古老教堂的外墙看不出来它在历史上的作用，因为其四周建筑胡乱搭建，毫无布局。如今，情况则大不相同。

为这一古老教堂建一个安全的屋顶要根据周围环境来合理设计。新设计的几何图形屋顶源于对原来屋顶的抽象处理。其次，建筑师通过使用合适的建筑结构材料和屋顶材料来保护墙壁和室内空间。在保证室内空间宽敞明亮的同时，材料也要持久耐用。最后，墙体要实现自然通风和照明。

通过多次与利益相关者进行对话、收集他们的要求，然后到实地

考察，建筑师最终确定了设计方案。设计方案考虑因素颇多，其中最重要的是以下两点。

——身份：这个新屋顶（大约330m^2）设计是对原来山墙形屋顶主要元素的抽象处理。通过分解这些元素，将其倒置，从而产生一种双曲面的效果，让人不禁想到原来的屋顶形状。双曲面平面的扭转旨在与周围静态环境形成鲜明对照，重现了这座古老教堂的价值。

——材料：建筑师建议使用市场上现有的材料，这些材料能够满足设计所要表现的建筑特色，另外又与原遗址结构区别开来；此外还使用了金属结构（由小型型钢组成的金属结构横跨整个屋顶），以及半透明的聚碳酸酯材料（在保证室内获得充足光线的同时保护室内空间，同时有助于增加屋顶平面的扭矩）。

Baños Community Center

In response to the need for recovery of Ecuador's sites listed as cultural heritage, the ministry, through the emergency management unit, hired the professional services for a new roof over the oldest church in Baños. These sites constitute a key factor for local and regional development.

The old church in Baños is located next to the current Municipality (dating from 1788). The church was repaired after the earthquake of 1797, but it was finally reduced to ruins by another earthquake in 1949.

The walls of the old church could not demonstrate the hitorical role until today as they were confined by several buildings chaotically laid out.

Making a roof for the old church that is safe needs an appropriate design for its environment, and geometric shapes are obtained from a process of abstraction of the original roof. Protecting the interior walls and space with structural and roof materials that ensure their durability to achieve an open interior space, and finally naturally ventilating and lighting the space confined by the walls are archieved.

The proposal was born from the collection of requirements generated from different conversations with the stakeholders and the conclusions obtained from the visit to the site. Among the many values, these two were especially considered important.

– Identity: The new roof (approximately 330m²) arises from the abstraction process of the main elements of the original gable. Decomposing these elements and facing inversely, they generate a hyperbola that virtually recalls the original roof.

The twisting of the plane that forms the hyperbola seeks to contrast with the static environment revaluing this ancient church.

– Materials: The architect proposed existing materials in the market that meet the necessary characteristics to adequately represent the design and differentiated from the original structure; besides metal structure(covering large spans with small sections), translucent polycarbonate(protecting without subtracting adequate lighting inside and facilitating the torque of the roof plane) is also used.

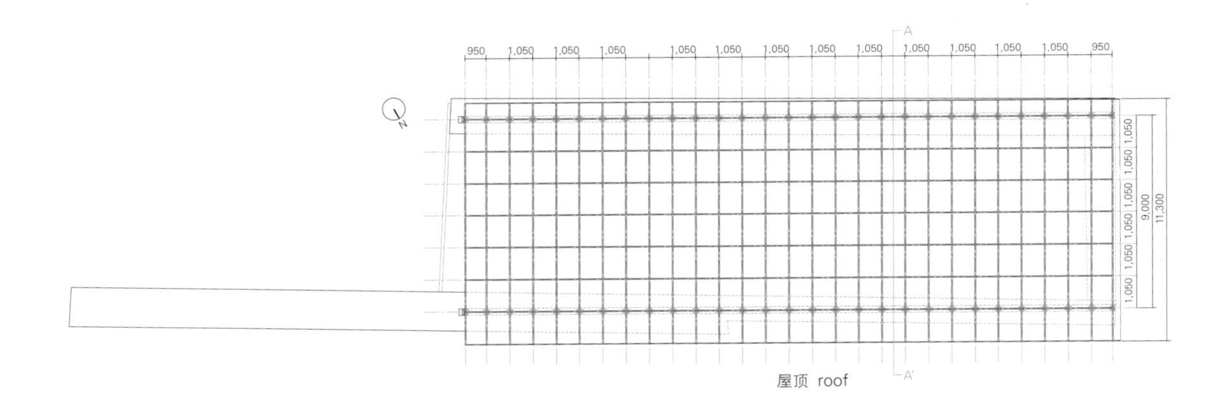

屋顶 roof

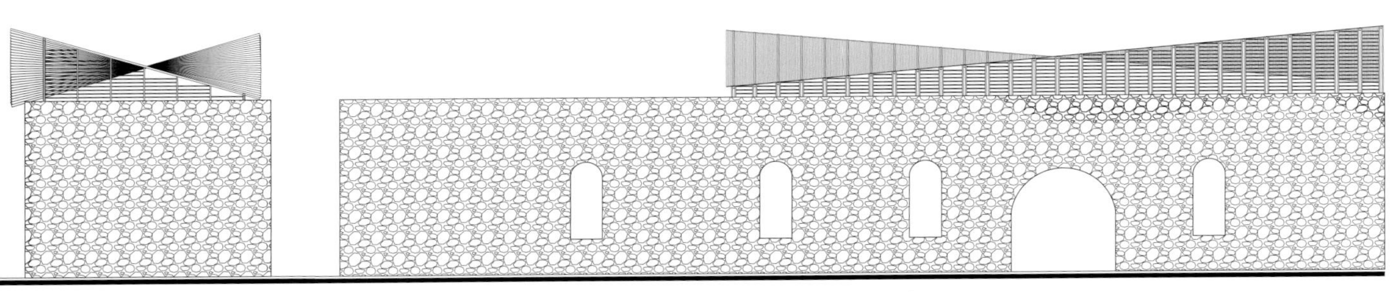

东立面 east elevation

北立面 north elevation

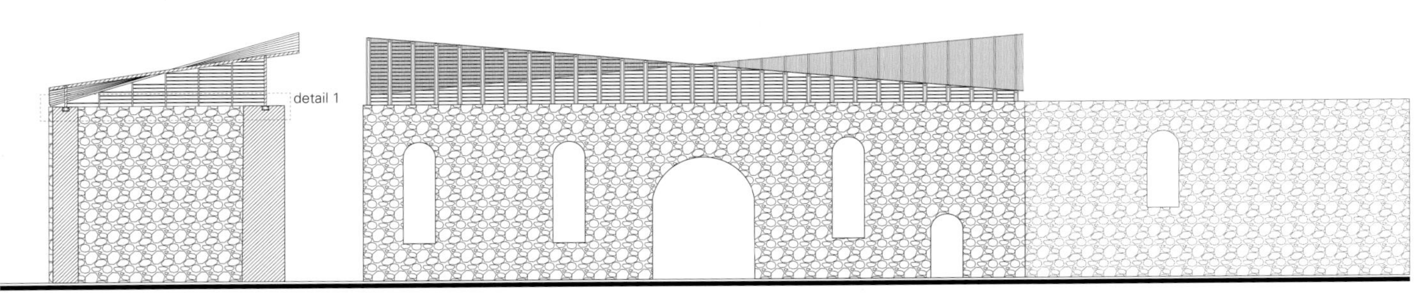

A–A' 剖面图 section A-A'

南立面 south elevation

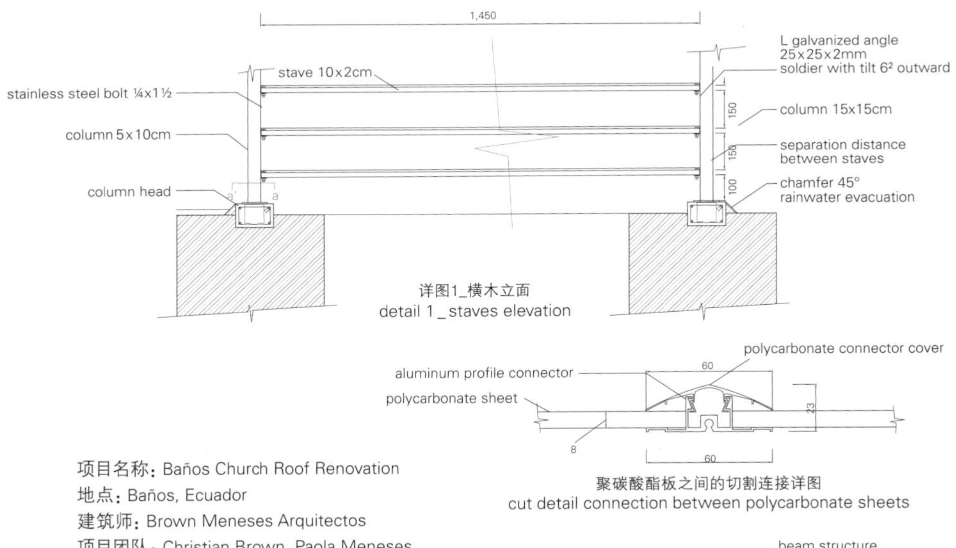

详图1_横木立面
detail 1_staves elevation

polycarbonate connector cover
aluminum profile connector
polycarbonate sheet
60
23
8
60

聚碳酸酯板之间的切割连接详图
cut detail connection between polycarbonate sheets

项目名称：Baños Church Roof Renovation
地点：Baños, Ecuador
建筑师：Brown Meneses Arquitectos
项目团队：Christian Brown, Paola Meneses
结构设计师：Ing. Félix Vaca
结构工程师：Sonia Perez
屋顶面积：437m²
设计时间：2009 / 施工时间：2009.5~10 / 竣工时间：2010
摄影师：Courtesy of the architect-p.136, p.137, p.140
©Sebastian Crespo (courtesy of the architect)
-p.138~139, p.141, p.142~143

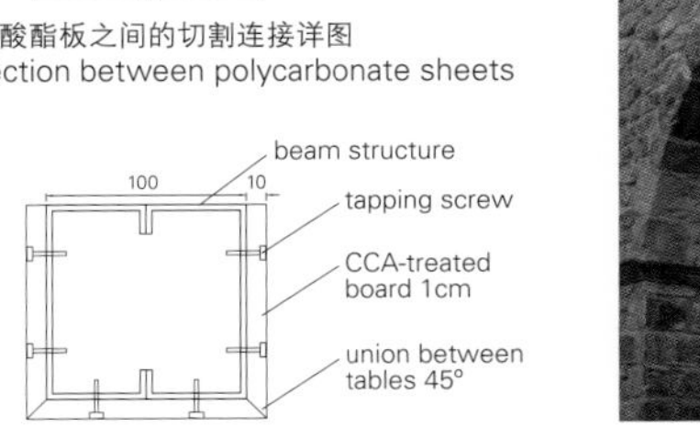

a–a' 剖面图_带木梁的详图
section a-a'_detail lined with wood beam